sufficient

Tom Petherick

sufficient

a modern guide to sustainable living

Illustrations by Lotte Oldfield
Photography by Melanie Eclare and Francesca Yorke

PAVILION

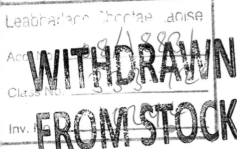
For Melanie, with love.

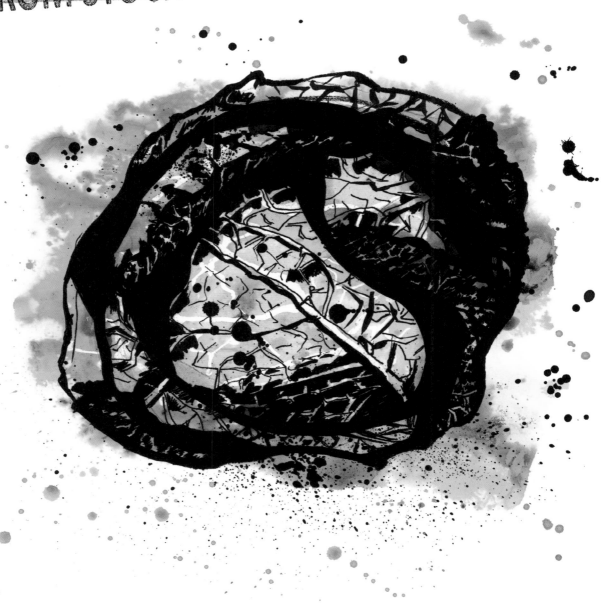

contents

INTRODUCTION

Most of us have enough. The title of this book is designed to make us look at the word 'sufficient' in two ways. It asks firstly that we become more 'sufficient' and responsible for our levels of consumption and secondly that we change to a more self-reliant way of living. It is about making the best possible use of our resources and potential, about an understanding that, amongst other things, working the land whilst appreciating nature is rewarding; it is a book about feeling satisfied with what we have – in short, 'sufficient'.

No one can doubt that we live in an era of mass overconsumption. This book has at its core the need for us in the developed world to accept that we need to make very significant changes. So comfortable have we become in the ways of convenience that sometimes it appears we cannot manage any other way, with the result that we consume even more.

This then presents an opportunity to look how we can become more self-reliant, particularly on the home-production front. There is little point in lingering on how badly wrong things have gone – the question is what we can do to effect change for ourselves and our community around us and even the greater community, the planet. The world we live in is a staggering place, filled with wonder and beauty. Nature provides in such extraordinary abundance that it is at times jaw-dropping. This will not disappear overnight, but there is much to do. We are at the beginning of an exciting time when our true worth will come to the fore.

Whilst trying not to lecture, I do feel the need to urge a sense of responsibility in our actions – indeed the word 'sufficient' reflects the willpower to say, 'That's enough.' By being comfortable with 'sufficient', we set an example, a political example that can only lead to action on the part of many. For years, I have been advocating the benefits of growing organic vegetables and I was a certified commercial organic grower, producing crop for market. I have watched carefully the way in which attitudes have slowly begun to change, not just in the realm of food but in an awareness of all things 'environmental'. The media tells us that we are at a crossroads, a turning point, that a paradigm shift is imminent. But most of us don't need any telling; we know, and we know that now is the time to act like never before.

I hope that this book will guide and encourage. It is largely based on my own experiences – certainly on the growing front – and it is a privilege to be given the chance to write about things as I see them and have experienced them. I have not provided an exhaustive catalogue of every crop or every technique, but have chosen those that I enjoy growing and eating, or which suit the 'sufficient' way of being. Naturally, readers will experience things differently in the course of their own gardening or growing lifetimes.

From a young age, I always preferred to be outdoors rather than in. My mother taught me to garden and also encouraged an interest in ornithology. My mother took my brother and I bird-watching to the wetlands of Norfolk to see avocets and stare out over the wide East Anglian skies. A reverence for nature was encouraged by my parents and has stayed with me throughout my adult life. We also had a vegetable patch in our garden and a large compost heap where Dad put the lawn mowings and raked up autumn leaves and Mum deposited the kitchen scraps. I remember the heat and the thick smell of compost, and I love the smell to this day. It is the bedrock of all gardening.

Today, more and more of us seem to be afraid of outdoors. It is an unfamiliar place dominated by that appalling variable – weather. Either too hot or too cold, most of the time wet, it never seems to suit us and prevents us from seeing outdoors for what it really is – life. The life of the air, the rain, the clouds and the ground beneath our feet. All of it is filled with life, and failure to embrace it is a terrible waste. It is where our food and the air we breathe comes from; without it, we are bereft of any connection to life.

Curiously, we spend a huge amount of time complaining that outdoors is becoming subsumed under a mass of roads and new housing developments whilst at the same time failing to make the best use of the outdoor space we have, from our wonderful National Parks to our back gardens. Most of the time we are staring our most precious resource in the face – that little patch outside the front door where this embracing of life and the natural world can actually begin.

Most of us have something that resembles a garden, or at least some access to outdoor growing space, even if it is just a balcony or a windowsill. Something as simple as growing radish from seed in a container can make the difference and encourage a new way of thinking. To connect with nature is the first step on the path of 'sufficient'. Long before I started to grow things, the bird table and the nut feeder were a source of endless fascination. The cycles of life and the seasons soon followed. Then came the soil and what it could provide.

When I was a young child in the 1960s, an attitude of 'sufficient' – particularly towards food – was prevalent in our family. Perhaps it was the legacy of my parents being children during the war years and experiencing food rationing first-hand, or it simply may have been the attitude of their generation and the fact that 'ready meals' were not available. If there was a cake needed, it got baked. We keep hearing today about the importance of returning to seasonal and local eating, but back then there was not much choice. It was all seasonal. Vegetables for city dwellers came largely from the market garden industry around London and fruit from Kent. How can it be that in only 40 years there has been such a decline in food production in the UK? The information age combined with cheap travel and changing tastes brought us a whole new world, and before we knew it the greengrocer was replaced by the supermarket. Most of this passed my family by; we grew a lot of what we ate, meat came from the butcher and it was only bananas and oranges that could be termed as 'exotics'.

In 1976, amidst all this, John Seymour's book *Self Sufficiency* was published and came into our house. At a stroke, all that I thought was great about 'outside' was suddenly OK. Seymour's classic became my bedside reading and subject for fantasy. Even the *Joy of Sex*, by Alex Comfort (same brown drawings and beards, though totally illicit reading) could not compete with *Self Sufficiency*, which was for me the life-changing book. I was utterly gripped, yearning only for that one-acre holding with all that it promised.

My first career after school was in horse racing, during which time I worked for trainers and stud men in the wide open sweeps of East Anglia, Wiltshire and Ireland. I found these wonderful landscapes impersonal, daunting even. The only contact between man and soil came through the use of huge machines. I yearned to see people growing crops and producing food in a way that I could understand and relate to. I started gardening properly myself when I was 20,

but it was not until I was 27 that I finally enrolled to study at horticultural college to take the business of growing seriously and learn a trade that had fascinated me since childhood.

Perhaps from Seymour's book or perhaps from my mother's example, I always felt that growing food should be done on a small scale, and in some ways I still believe this today. A big part of this book is concerned with trying to encourage a connection with the land, which we have lost through any number of reasons beyond mechanisation and industrialised farming. While we may not be able to grow our food or comprehend where it comes from, we might start by going out into the landscape and looking for wild food or finding a farm that sells produce at the gate. It is a first step, and one that our children are never likely to forget. We all have memories of childhood, but the good ones stay with us, stand out and usually ring true. As a child, I spent a lot of time by the sea in Cornwall. My father taught me to fish and catch shrimps, which I still do today. While the county's horticultural history is famous, it was even more significant those 40 years ago. The market garden industry in the Tamar Valley, which separates Cornwall from Devon, was flourishing. There were varieties of apple growing in Cornwall that never made it to London. Thankfully, this local industry is on the rise again, but it was the vibrancy of the local and seasonal food, to which I could have some connection, that spurred me on to maintain a passion for growing.

As the world around us seems to scale up and enlarge (globalise and centralise), it is surely time for those of us who feel the need to be more 'sufficient' to scale down. The Stern Report considering the economic effects of climate change (first published at the end of 2006 by the Britsh Treasury) made many incisive judgements about what might happen in the future. Tourism would be hit badly as air taxes rise, energy costs would spiral and, generally, life would become less 'convenient'. Although the report was long overdue, it neglected to say two important things that concern us in this book: that firstly we need to scale down our consumption and secondly we need to learn to be more self-reliant, and less dependent on outside resources. That way, we can make a difference, and possibly even a living, because scaling down and being more sufficient is a new way of thinking that will inevitably create its own energy, and from energy come resources and from resources come money.

A less consumptive way of living is no longer an idea confined to the old hippies amongst us. It seems a proper way to live. An entirely new set of personal rules are needed. Money will not be able to buy resources, because they will not be there for us to buy. Last year, for example, the UK's supply of gas nearly ran out. That would have been interesting! We must teach ourselves how we are going to cope in the future – and that future is not way in the distance; it is now and we must act.

So how do we set out on this path? How do we say no to that new car? What else will make us sit up and listen? What will make us change to the green energy supplier or put solar panels on the roof? Is it conscience, a belief that we might save money – or are we going to leave it to our children to make these decisions when it is too late? There is no more time to waste. This is real. We live in the days of peak oil, and the finite resources that we have taken for granted are in their death throes.

My feeling is that we will get it sooner rather than later. In the 20 years that I have been advocating organic food, there has been a sea change. The fact that 70 per cent of the organic food consumed in this country is imported says little about our desire to eat organic and more about the state of our farming industry. But those food miles and the absurd system of farming in the European Community are unsustainable. If we are going to be able to source what we want, or better still grow it, or encourage our neighbours or farmers to, we are going to have to do it ourselves and persuade governments to support us and the farmers.

For a great many people, the thought of global warming and climate change is unnerving, whether they fear the danger that comes from living in a low-lying area at risk from flooding or are a business affected by an energy tax. To my mind, this is a chance for a new beginning, and one that we must embrace. My fervent hope is that it will bring people and communities together to seek solutions. We have to take responsibility for our actions and tread that little bit more gently on the planet.

The truth is, we find living together difficult. We are not good at community. It is only at times of disaster that we hear of communities 'pulling together'. Communal living – an idea that sprang up in the 1960s – never really works because, as one long-serving member of a community once told me, 'of each individual's failure to do the washing-up'. In short, we are not prepared to make sacrifices. Even today there are only a handful of really successful communities and eco-villages around the world, including Findhorn in Scotland, Auroville in south India, Bedzed in London, and Christiania in Copenhagen.

Perhaps we should try and focus on the more easily attainable and positive benefits of becoming more 'sufficient' and how good it feels to grow some of our own fruit and become less reliant on a supermarket. And, even better still, to make a contribution to local economy. Even the slightest gesture, such as buying a locally baked loaf at a farmer's market, brings us closer in touch with the local community, for everything and everyone in a tight radius is interlinked.

To me, the concept of working for the common good – the cause, if you will – has unending appeal. I am no saint leaving only fresh air as a carbon footprint, but I have been a commercial grower of organic fruit and vegetables and I saw how much that meant to people. (And that was in the 1990s, when organic food was still in its infancy.) I see the enthusiasm and almost childlike excitement that stirs in people who believe they are operating outside the system, individual, honest and alive. Prepared to experiment with ideas and practices that have low impact. It may sound worthy but it's worthwhile because we are important – and so is the planet.

WHERE ARE WE NOW?

WHERE OUR FOOD COMES FROM

Since agriculture proper began in the Fertile Crescent of Mesopotamia some 8,000 years BC, farmers and growers have been providing us with our food. They should be lauded for their efforts. To work the land is a wonderful thing, to bring forth healthy crops from soil in adverse conditions is no easy task. And no matter what technology we have available in this, the modern farming era of advanced agricultural mechanisation, we, like the Mesopotamians, still cannot control the weather and its affect on our crop yields.

In the 1950s, Philip Oyler called his book about life in the farming communities of the Dordogne Valley in southern France *This Generous Earth.* No one has put it better. Our soils want to grow plants; this is the natural response of most of the dry land areas of the planet, barring mountain-tops and deserts. It is our job, as citizens of the earth and consumers of the food that the earth provides, to see it is done properly. Curiously, the guidelines for care of the soil have remained virtually the same since those first farmers worked out how to keep their land healthy and productive. Whether these guidelines are followed, and the consequences if they are ignored, will be discussed throughout the course of this book.

Today, more and more consumers want to know the provenance of the food they are eating. Food labelling rules are, if not stringent, becoming tougher. 'Traceability', the buzz word that surrounds food, is as much about knowing where the product comes from as about being able to find the source if something goes wrong. And it has been going more and more wrong in the last decade. In the UK alone, there have been debilitating outbreaks of disease, including Mad Cow Disease, Foot and Mouth, Swine Fever and Avian Flu. As if there was not enough threat to world health from HIV and malaria, now food presents a danger. Food – how can this not be sacrosanct?

There are, of course, no end of so-called food 'scares' that amount to absolutely nothing and are the result of living in an age where we are so risk-averse. We have little chance of raising our immunity and becoming less susceptible to disease beccause of our obssession with hygiene and lack of exposure to germs. Nonetheless the big ones are here, and are almost certainly the result of intensive farming, where animals are reared in poor

conditions and fed an unsuitable diet. Meantime, the soil plays little direct part in the rearing of animals, whose health is affected because most of them are raised indoors.

So, we need to know where our food is coming from, and whilst we are consistently reassured by the food standards authorities that the food we eat is safe because their science says so, nature still finds a way around their regulations to spread and mutate disease. The irony is that this obsession with standards of hygiene, allied with greed, is probably the cause of so many of the problems in the first place.

Feeding animals to animals is merely a short cut, an economic choice borne of ignorance and greed. But farmers have to survive like everyone else, and their remit to produce clean wholesome food on large acreages using lots of chemicals is not easy.

In my early twenties, I spent some time living in Wiltshire, an area dominated by huge fields growing predominately wheat and barley. It was a beautiful part of the world, where agricultural plains merged with rough, uncultivated down land. I found it daunting and faintly unnerving that wheat and barley were grown as monocrops covering vast areas. This monoculture does not happen in nature, where diversity of plant material maintains a balance and acts as safety mechanism against the build-up of pests and diseases. In a monoculture, pest and disease can spread quickly, so the plants become dependent on artificial chemicals as a means of defence. Also, without animals in the system, it becomes difficult to return organic matter and nutrients to the soil, so the plants have to be artificially fed. When this cycle began, we started down the rocky road of ruin.

In his excellent book *We Want Real Food*, Graham Harvey, a true expert on the subjects of farming and nutrition, says that we are 'fifty years into a mass experiment into human nutrition. We are all eating basic foods that have been stripped of the antioxidants, trace elements and essential fatty acids that once promoted good health.' 'Is it any wonder,' he goes on, 'that our body maintenance systems are breaking down in middle age or earlier?'

Up until the 'Green Revolution' of the post Second World War era, it was a given that food was healthy. The foundations of soil fertility – rotation of crops and the returning of animal wastes and other organic matter to the soil – were strictly followed. No-one gave any thought to 'food scares' but had a healthy respect and knowledge of how to store and prepare fresh food in

the home, as there was little in the way of refrigeration. Undeniably, there was far less choice, but the product, whether meat, dairy, fruit or vegetables, was sturdy and life-giving because it was grown in strong healthy soil.

The age of chemical farming developed quickly. Its intention was quite simply to feed more people in the world and, on the face of it, this was a good idea. Countries in the developing world soon had access to cheap chemicals, which were intended to realise bumper crops to increase their poor levels of production and feed their rapidly growing populations. Feeding an increase in population will probably be the biggest challenge that we will all have to face in the future, especially as agriculture yields are directly affected by changing weather patterns due to climate change.

At first, the results of the 'Green Revolution' in South America and Asia were indeed startling, with crop yields increasing. But farming on a grandscale with chemicals is a bit like 'slash and burn' agriculture – the process of burning areas of a forest to clear it and make way for crops. The soil, enriched by the ash from the fire, is productive for a few years and then, with nothing going back in to the system, the soil becomes denuded, the crop fails and the 'slash and burn' farmers move on.

In the developed world, a wholesale uptake of chemical farming and a move away from the old routines followed the post-war 'Green Revolution'. For conventional crops, this had a disastrous effect and has created a greater reliance on chemicals than ever before. The soil is so utterly worn out that it now merely acts as a rooting medium. The soil does not nourish plants any more. The plants are entirely dependent on chemicals to grow. And still we hear that there is no evidence that organically grown crops are healthier than conventional crops. In my view, people who deny the benefits of organic methods are up there with those who deny climate change is happening.

And this, in a nutshell, is where our food comes from. The purpose of this book is not to try and solve world hunger, it is to empower and educate the reader about food. Even if you are unable to grow fruit or vegetables yourselves – although it is my sincere hope that some of you will try to grow something – this knowledge will help you make informed choices about the food you buy and help you discover its provenance.

If conventional agriculture with its heavy reliance on artificial and fossil-fuel dependent chemicals does not appeal, then what are the alternatives? Clearly there is farming without chemicals in the old-fashioned manner, but

Left: There are no straight lines in nature. Contrast this fruitful, diverse potager garden with a commercial monocrop.

then there is the word 'organic'. How did this arise? What does it really mean? And does it work? I am convinced that the organic ways of farming are the right ways, but Masanobu Fukuoka, the Japanese natural farmer par excellence, says organic farming is simply another form of scientific farming with materials moving 'first here, then there, to process and treat'. We are still interfering with nature and not allowing her to redress the upset in balance caused by farmers, be they organic or otherwise.

The term 'organic farming' was first used by Lord Northbourne. He began using it in the 1930s to describe the type of holistic, balanced agriculture that he was practising on his farm in Kent in southern England. The other early pioneers of organic farming, such as Lady Eve Balfour and Sir Albert Howard in the UK, the grocer's son J. I. Rodale in the US, and the Austrian founder of the bio-dynamic movement in Europe, Rudolf Steiner, also went into print in support of organic farming. Howard wrote prolifically about his experiments with compost in India between the wars, while in the 1940s, Balfour scientifically compared conventional to organic agriculture on her farm in Suffolk in a process that became known as the Haughley Experiment. In 1961, Rodale published *How to Grow Vegetables and Fruits by the Organic Method*, which became the best-known book on organics for the American home-gardener.

The extraordinary thing to me is that these early pioneers, whose books radiate sense through good husbandry and science, failed in their bid to convert the masses to organic farming and growing. It is only now, in 2007, that we are beginning to embrace organic food with some intent. But still in the UK, the total area under organic cultivation is just over 3 per cent of the total farm land. Quite some way to go. (But statistics never tell the full story, for there are people whose have never registered their land as organic but have always grown food without chemicals.)

In the UK, we still import around 70 per cent of our organic produce. For a nation of farmers bound by a temperate climate and more than adequate rainfall, we cannot bring ourselves to move wholesale into organic farming. As such, the vast majority of the food that we eat is grown by conventional methods while organic methods lag sadly behind. They are catching up, but very, very slowly. This means that most of our food is grown with the use of pesticides, herbicides, fungicides and artificial fertiliser. Many of these are toxic and non-selective. Worryingly, there is also the threat of genetically modified crops creeping closer despite public alarm and, in some cases, civil unrest.

The organic market is growing each year to the tune of hundreds of millions of pounds. Sadly, attached to this is the belief thrust upon us by politicians and the media that eating organic is a 'lifestyle choice', as though the choosing of organic food over conventional food says something about who we are. I would have said it has more to do with common sense, particularly in light of the fact that everyone knows by now of the appalling results of chemical farming. It is the way of the world that vested interests, especially the farming lobby, are too strong to allow governments to upset the equilibrium. The organic movement is simply not robust enough to take on the might of the agrochemical industry. This is sad because organic practices do work; they work on behalf of nature and the good health of the consumer.

THE STATE OF AGRICULTURE AND HORTICULTURE TODAY

Twenty years ago, when I started out in growing, I was able to go to college and take a three-year National Diploma in Horticulture. The college had been founded in the 1920s with the purpose of providing the glasshouse industry of the Lee Valley, just to the north of London, with trained growers. In the 1920s, those thriving glasshouses supplied the London fruit and vegetable markets with tomatoes and cucumbers. Today, more so than ever, this localised industry no longer exists, and fresh produce sold at New Covent Garden market is flown in from all over the world.

The UK's once-great horticultural industry lies virtually in ruins because gardeners cannot get the proper practical training required in the cultivation of plants. There are still many colleges of horticulture in the country but, such is the state of the industry, that there are more courses covering the theory of garden design, sports turf management, equestrian activities and floristry than in learning the practical methods of growing edible and ornamental plants.

The organic question and the failure of industry and mainstream agricultural colleges to embrace it has left a huge hole. All those years ago at college, when it came to my final examinations, I informed my course tutor that I was going to answer all my exam questions from an organic perspective and that when questioned about chemicals for treating pests and diseases, I would offer organic alternatives. Without hesitation, my tutor told me I would be failed for such a heinous crime against the Establishment. I called his bluff and passed, but such was the blinkered attitude to change that I was more than a little gloomy when I set out from college on a 'chemical-free' crusade.

It was only a couple of years after I left college that apple growers in the UK were going out of business because of the spiralling costs of chemicals. They refused to switch to organic methods and try different varieties of apples, so many lost their businesses – as farmers are doing all over the world at this very minute in the face of rampant opposition to anything organic or sustainable. The UK government would rather farming did not exist at all since, they claim, it would be cheaper to import everything we eat. In the US, if you are a not a giant farmer embracing GM technology, no one wants anything to do with you. The only money available for experiment and

Right: The impersonal landscape perfectly defined by a monocrop of wheat.

research in the agricultural industry is for GM, not organic. Governments across the globe talk forever about sustainability, but fail to see that agriculture is the area where it matters most.

Some years ago, I took over some land to grow organic crops commercially. I registered myself with the UK's premier organic certification body, had my land inspected and made a start. It was in Dorset, a rural, sparsely populated area of southern England with an emphasis on agriculture rather than people, a place wholly unaffected by trends. At the time, it was what I wanted to do more than anything, and I used it to trial crops and markets, and as a sounding board for the state of the industry in general. I was right about organic cultivation, I knew in my heart of hearts that this was the way food production had to go. Trying to persuade everyone else had been nigh on impossible, for they just would not have it, until I began coming up with the product. Then things changed.

My crops were primarily salads with some spinach and beans to put nitrogen back into the rotation system, which the certification body rightly insisted upon. I did not run an organic vegetable delivery 'box scheme'. Instead, I delivered fresh salad to a couple of shops, one or two friends and a pub whose landlord was an enlightened chef who knew about the importance of good, local ingredients. Once a month, I took my produce to a farmer's market, where I encountered the general public. This was where it got interesting. The response to the produce was phenomenal and I rarely had difficulty in selling out; most weeks, I could have sold at least twice what I took with me. I sold to a wide cross-section of people, but their requirements were the same: fresh food that had been grown in healthy soil. If that is the definition of organic, then that is what I had, and they loved it.

While farmer's markets are booming, many farmers who sell to the supermarkets are struggling. The primary reason is that often farmers find themselves selling in product at little more than it costs them to produce it, such is the might of the giant retailers.

So, the farmer has to diversify to survive. Converting buildings for different uses, switching to organics and learning to grow something else are becoming the only real options. Not so easy for the hill sheep farmer in the teeth of a northerly gale! As consumer's shopping habits change, more trade is starting to be done in local produce. There is a sea change afoot and future generations will hopefully know nothing else other than proper food.

Right: The organic vegetable box; a growing trend in fruit and vegetable retailing.

SUFFICIENT FOOD

Whilst I try and put the case in favour of food grown without chemicals and artificial inputs, and indeed have been a certified organic grower, this does not mean that everyone should follow suit. Naturally, I hope that this type of cultivation appeals to growers, particularly first-time gardeners, because there is nothing to lose, especially if you know no other way.

I feel that the concept of 'sufficient' goes hand-in-hand with nature and natural systems. Nature is an excellent guide, for she only ever does enough, never offers more than is required, and on her own, without interference, is in a state of balance. It is this state of diversity, balance and minimal interference that we should try and replicate when it comes to growing things. And it is why the followers of the revolutionary agriculturalist Masanobu Fukuoka appeal for the natural way of farming – no tillage, no fertiliser, no pesticides, no weeding and no pruning.

I like this 'hands-off' approach when it is applied to the right area of the garden – an orchard, perhaps – but I would not recommend it as a system of gardening for a small space that needs to be well-ordered and maintained. As an experiment, I have let a polytunnel go native for several years to see what would happen. I have even stopped watering and don't even bother to remove the dead plant matter. What appears is astonishing. One or two plants have become dominant, in particular, self-seeders, such as borage and nasturtiums. And up through the wilds and the weeds come seedlings of all kinds of plants that once grew there and others that I don't remember planting. Perhaps the artificial environment of a polytunnel is not a fair reflection of Fukuoka's permacultural, self-seeding, wild aspect of gardening. However, I think his methods are perhaps better suited to hotter climates than our own cool temperate weather pattern, as you naturally get greater bio-diversity in a warm climate.

All that said, you have to start somewhere and get planting to see if you can grow anything at all. A more complex, 'hands-off' system takes some understanding, and will probably follow on as your knowledge increases. But as I point out repeatedly in *Chapter 2: Growing*, it is easy to grow too much and quickly denude the soil of its nutrients. This is the 'Oliver Twist guide to gardening': if you ask for more, you may well get rebuked; if you settle for 'sufficient', all will be well.

Just remember that whatever you take out of the ground in the way of a crop you must put back by means of nutrition. Lack of food for the soil is the most common mistake made in gardening. Whilst nature has her own means of fertilising ground, she cannot compete if you repeatedly rob the ground of its 'life force'. Everything in the natural world is cyclical, and the cycle will break if you do not put back what you took out. It is essential that you go about your growing with this attitude. Accept what comes, enjoy the experience, observe and intervene only when necessary. When plants need water or support, water them or offer them a stake to hold them up, but remember that nature will ultimately find her own way, even though it was you that planted the seeds.

SLOW FOOD

No organisation is so closely aligned with the ideas of 'sufficient' as the Slow Food movement. This organisation, which came into being in Italy in the 1980s, was borne out of a realisation that what it calls 'fast life' was resulting in bad food – badly grown and badly consumed – and that this was also having disastrous consequences for the environment. Allegedly one of the turning points in the foundation of the movement was when a well-known fast food chain opened within earshot of the Coliseum in Rome.

The Slow Food movement has at its heart the realisation that farmers, especially niche farmers with perhaps not much land or financial stability, deserve to be recognised for the product they produce and to be properly paid. The movement came up with the idea of calling its members 'co-producers' because it felt that by supporting the grower and producer (by purchasing the product), the consumer was inevitably involved in the process.

This communication between the people who produce and consume the food would lead to a clearer understanding of what happens in the world of food. I found this to be true when I ran my organic market garden. In the shops I supplied, I was told that customers wanted to know more about how and where the food was produced, as they did when I sold direct to the customer at the farmer's market. On my stall, I was questioned at length on varieties of lettuce and planting spaces for asparagus crowns. As I got to know my regular customers, they would come and visit me in the working garden, furthering discussions about good food and proper farming practices. Real community stuff.

The mission of the Slow Food movement is as follows: 'Slow Food works to defend biodiversity in our food supply, spread taste education and connect producers of excellent foods with co-producers through events and initiatives.' There are currently 80,000 members worldwide, who are attached to 850 different convivia, or local chapters. It is the biggest and most international organisation of its kind and links thousands of producers and co-producers across the world. Its festivals of food are without compare, where the lesser known countries link with the giants to encourage a global food consciousness.

Above and right: Vegetables and cheese on sale at the organic market on Boulevard Raspail, Paris, France. Markets such as these, where producers of top quality food come together, are becoming very popular.

SOURCING AND PROCURING

Hopefully, your intention is to grow as much food as possible yourself, but as I point out in later chapters, there are some things which you may find impractical to grow for reasons of space, expertise, particular pest problems, shade or whatever. This is where some knowledge of sourcing and procuring local fresh ingredients can become useful.

Thankfully, sourcing and procuring food is becoming easier by the day as local food directories list an ever-increasing number of producers and growers. Many of them offer home delivery services, which are more convenient for the consumer and also the best way to reduce fuel consumption, as they deliver to several addresses in the course of a single journey.

Other options include doing 'straight swaps' with friend or neighbours for produce. Enter into an arrangement with a neighbour whereby you grow the salad rocket and they grow the hay. I love growing cucumbers, but I cannot seem to grow aubergines with any success. Perhaps over the fence lives the county's aubergine specialist? Communication is everything and gardeners are a cheery and generous lot by definition, happy to swap information and trade secrets. Get to know as many of them in the community as possible, and make swapping fresh produce a truly viable option.

Also, there are organic box delivery schemes, which offer high quality vegetables in season, delivered to your door. Try to enter into an arrangement with the grower to make sure that you get what you want in the box rather than just what he has available. For example, in the depths of winter when fresh greens are in short supply, try to make sure that you don't just get a box each week full of roots, potatoes and squash.

Then there is the possibility of entering into some other sort of arrangement with the grower whereby they provide you with certain products on a regular basis but not within the limitations of a box scheme. This is often done on a contract basis, where you guarantee your business for a certain length of time and they guarantee to grow and supply certain crops for you.

This leads on to Community Supported Agriculture, which I think is the way forward for those of us who are unable to grow our own. Like the Slow Food movement, it allows you the consumer to take an active role in the farming process because effectively you are buying shares in the farm. Community Supported Agriculture is taking its time to get going on account of the considerable investment required, both physical as well as financial. (By this I mean, the farmer first has to make the big financial decision to change his methods of production. Also, starting up a new project from scratch, takes time and energy to bring together the right people and resources.)

The great thing about the system is that the farms involved produce a wide range of crops beyond vegetables, invariably fruit, meat and dairy. They are almost always organic or bio-dynamic and they have very strong links with the community, which is not always the case with a small individual business, such as a box scheme. Community Supported farms tend to be the hub of that community and while the movement is further advanced in the US, the model is being taken up in Europe with some purpose.

My experience of Community Supported Agriculture is not extensive, but I did spend a couple of days working on one scheme, called Earthshare, in north-east Scotland. Here, as I suspected, I found energy and compassion being actively poured into the project. The love of and respect for the land by the people involved was clearly evident and I spent two very happy days enjoying the feeling of well-being that comes from helping to produce clean food for people who genuinely appreciate it.

Closer to home are the farmer's markets, which happen with increasing regularity, especially during the growing seasons. Again, the range of produce is wide, it is guaranteed to be fresh and grown within a certain radius of the market. By shopping in this environment, you can get to know your local growers and producers. Farm shops, attached to a farm where at least some of the produce is grown, are also worth investigating. Find out where they source the produce that is bought in so you know its provenance.

Above all, it is important to make a connection with the food you are eating and to have a good understanding of its nutritional value. It is not enough to say, 'Because I eat organic, I am getting what I need.' For some, buying food direct from an organic farm, whether over the counter or on the Internet, is enough reassurance. But if you make a direct connection with the farm and the animals and have some knowledge of their system of growing, you will absolutely know for sure what you are getting in the way of nutrients.

Nutrition

Although this book is primarily about strategies for becoming more 'sufficient' in our lives, you need to know what is good for you nutritionally and understand what the human body needs to keep it in optimum health.

The great heroes of the modern organic age understood this requirement well. In his ground-breaking book *An Agricultural Testament*, Sir Albert Howard, the 'Godfather of organics', asks how it is possible for anyone to ascertain the connection between a fertile soil and the health of the people who consume the crops or animals produced on the land. Howard sought out the evidence raised by the nutritionist Sir Robert McCarrison to find answers.

From his research in India, McCarrison noticed 'suggestive differences' in the physique of the various races. The physique of the northern races, he said, was strikingly superior to that of the south, eastern and western tracts. McCarrison found that the causes of the differences corresponded directly to the food the people ate.

In northern India, and what is now Pakistan, the diet consisted mainly of wheat, but all the proteins, vitamins, minerals and salts in the grain were eaten. After that came milk and its by-products and then pulses, vegetables and fruits. Meat was eaten only sparingly. In terms of diet, little has changed today amongst the hill tribes of the north-west frontier since this research was carried out in the 1930s and 1940s.

In the east, west and south of India, the main staple is rice which is parboiled and polished before being cooked and eaten. This highly refined rice has almost no goodness left in it by the time that process has finished. All the nutrients have been removed. Also, not much is eaten in the way of milk or vegetables and fruit. Meat to this day is very rare. Amongst the rural populations, most meals consist of massive quantities of rice with tiny portions of vegetable, some pulses and very occasionally an egg. The indigenous people who live on this diet of polished rice are short in stature, in comparison to the sturdy, strong native people of the north.

The people in northern Pakistan (whom McCarrison cited as being the finest race of all), worked their ground well, being careful to return all animal and human wastes as well as compost to the soil. Their mountainside soil was thin and much in need of nutrition. Also, as Graham Harvey points out in *We Want Real Food*, the terraced beds of the mountainous north were regularly soaked in 'glacial run-off', which is rich in trace elements, limestone and volcanic elements. With proper attention paid to soil nutrition, the result was superior human nutrition and an almost entire lack of disease.

The subject of western nutrition is addressed in Sally Fallon's book *Nourishing Traditions*. She asks why it is that even though millions of Americans have adopted a low sugar/salt diet, with fewer saturated fats, fresh fruit and vegetables and regular exercise, they are still falling victim to degenerative diseases, when only a generation ago these sorts of diseases were very rare. Her answer is that despite the recommendations of what she calls 'the diet dictocrats', no distinction is made between 'the traditional foods that nourished our ancestors and newfangled products that are now dominating the modern marketplace'.

Sally Fallon is the founder of the Weston A. Price Foundation of Washington, DC, a body dedicated to education, research and activism in the field of nutrition and food production. In the 1930s, Weston Price spent years studying the farming traditions and food consumption of indigenous peoples. He became convinced, like Howard and McCarrison, that traditional farming practices and diets held the keys to human health. This US Foundation promotes farming and growing as it has always been done, and so should we.

In the developed world, we eat food from all over the globe and from every ocean. Our diet is as varied as it possibly can be, but somehow that luxury has not afforded us the health that it should. I believe that one of the few ways in which we can come by food that will nourish us properly is to grow it ourselves in soil we have nourished ourselves.

FOOD MILES AND FAIR TRADE

The issues surrounding how far our food travels and how well the producer is paid are sensitive. A recent United Nations study shows that the majority of food miles are used by consumers travelling in their cars to buy food rather than by food being transported by air or sea to arrive fresh at our supermarkets. The misconceptions about food miles and the use of fossil fuels in the importing and exporting of produce, together with the gross inequality surrounding free trade, are having a devastating effect.

Imagine for a moment, a grower of a product in sub-saharan Africa – and there are plenty, particularly of medicinal crops. He wishes to export his crop or product. He finds a market abroad, only to discover that his potential customer is unwilling to buy the product because it has to travel all those miles by air. Is this not exactly the type of grower we should be protecting? The might of the Western free market would suggest otherwise, and here we must find the balance between food miles and free trade.

I also find myself taking issue with those who would say that edible crops from other Continents should be passed over in favour of those produced at home. Well, maybe, but something in me says that the economies of the developing world need supporting. In the UK, for example, we have regular supplies of green beans sold into our supermarkets from African countries in winter. They are organically grown and certified as such. For me, this is enough. The fact that they are organic indicates the farmer has probably made some serious sacrifices to get his crop to market. He has learned to farm his land in harmony with nature and, what is more, he has probably always farmed this way. Care of the land and of nature and a willingness to work in harmony with it is as important as the food miles of the righteous.

Products that are fairly traded through different 'Fair Trade' schemes may also have to travel some distances to market. But here we move towards the ethical side of trade in that the grower or producer is offered proper rates for the product and is further compensated for his troubles by receiving technical help from the fair trade organisation. The products most often in question are tea, coffee, bananas and cotton, but we can see that these are all consumed in quantity in the West, and that the producers are aided in their efforts is greatly to the credit of the Fair Trade system.

Right: The giant retailers have made life very hard for large- and small-scale producers alike. This is no way to procure food.

THE WAYS AND MEANS

The practical minutiae of what and what not to grow will be dealt with in *Chapter 2: Growing*, but what wider ranging systems or examples exist to help us on the path to 'sufficient'? Who and what can guide us in our efforts towards self-empowerment in this field?

I have already nailed my colours to the mast as regards organic cultivation, having practised it for many years. Nothing about it has persuaded me that it is not an excellent way to improve soil fertility and to grow healthy plants. It concentrates on the soil, the *Living Soil* as Lady Eve Balfour called it, the lifeblood of the garden and of human health. The organic movement is far-reaching and the same general principles of crop rotation, composting and fertilisation from animal and plant wastes have been in place for thousands of years. It is only since the advent of agrochemicals that we have lost our way. A mass return to organic cultivation is inevitable as the need to preserve our precious resources of non-renewable fossil fuels gets stronger and more imperative. The 'artificials' – fertilisers, pesticides, fungicides and herbicides – require huge energy consumption for their manufacture, and this will have to cease. As far as genetically modified organisms are concerned, their inexorable rise is a huge threat, being more anti-nature than any form of conventional chemical-based agriculture.

But while organics is an excellent method of cultivation, where else can we find guidance on how to make further steps towards a more 'sufficient' way of being? Is there a road map, a timetable, a set of rules and regulations? Do we need such a thing anyway? If a desire to work with nature is there, will most of the actions we take come naturally?

Some of those that have influenced me over the years include Masanobu Fukuoka, who wrote the following about the realisation that changed his life in his magnificent *One Straw Revolution* (originally published in Japan as *Shizen Noho Wara Ippon No Kakumei*): 'Humanity knows nothing at all. There is no intrinsic value in anything and every action is a futile, meaningless effort'.

This extraordinary Japanese farmer decided to put those thoughts into the practice of farming, growing rice and winter grain. The results were astonishing, and he wrote about them further in *The Natural Way of Farming*. His was a radical system of total non-intervention: no fertilising, no weeding, no cultivation and no pruning. He let nature take its course. Few have dared to tread where Fukuoka led, but this system has the ultimate respect for nature, and Fukuoka is compulsive reading. The path to 'sufficient' is wholly dependent on being intertwined with the realms of nature.

In *Farms of Tomorrow Revisited*, their sequel to the inspirational *Farms of Tomorrow: Community Supported Farms, Farm Supported Communities*, Steve McFadden and Trauger Groh describe three basic rules they learned from a lawyer friend who, they felt, had a deep understanding of the co-operative process of making such farms work. The first was not to work too many hours, the second involved avoiding expensive external inputs, such as machinery, but the third rule has the most resonance: 'Take all the initiative for your actions on the farm out of the realm of the spirit, not out of the realm of money'. Explaining what this means the lawyer friend goes on, 'The more we penetrate the spheres of nature, the more we become aware that what surrounds us in it is of an overwhelming wisdom. What we call scientifically an ecosystem is penetrated by wisdom so that all parts serve the whole in the most economic way'. The rule finishes with the understanding that the way a beehive is organised or the way that the soil life builds fertility by itself is what we can describe as the spirit that is spread out in nature.

Stirring stuff, and though Fukuoka's natural way of farming may be beyond most of us, we would do well to take that wisdom of nature with us on our journey. Turning our backs on nature has thus far led only to disaster. This is something that I have talked about while discussing Community Supported Agriculture (see pages 34–35), which was was born of the biodynamic movement founded by Rudolf Steiner in the first half of the twentieth century. Steiner was responding to a plea for help from local farmers in East Germany. Soils, they said, were becoming depleted from the use of chemical fertilisers, and animal and crop health was suffering.

Steiner gave a course of eight lectures in 1924, outlining his ideas for the future of agriculture. Biodynamics, said Steiner, should be an ecological system of farming but should attempt to achieve balance between the physical and non-physical realms. By 1928 a certification scheme, with the trademark 'Demeter', had already been set up for biodynamic foods. This is no ordinary way of farming; it is extremely complex and one that requires considerable effort and understanding. To begin with, just as the organic farmer sees the soil as a living organism so the biodynamic farmer sees the entire farm as one organism with everything in and around it interconnected: soil, water,

where are we now?

animals, people, crops, wild plants, insects and even the climate, the seasons and their rhythms.

But Steiner's spiritual/scientific take on agriculture encompasses more than an understanding of nature and her rhythms. What quickly developed was a process of composting involving a number of vital herbs sprayed onto the compost and onto the land in various preparations at critical times and moments during the seasons to activate both the compost heaps and the land. And all this in close connection with lunar cycles and planetary influence.

While we should understand the concepts of the various different systems or models that are in place, it is up to the individual whether or not they choose to practise them. Gardening, growing and developing with nature is a process that continually and gradually unfolds. We are in a constant state of learning and perhaps by selecting one specific system over another we limit our horizons; that the process should be free of chemicals has no bearing on this, for chemicals are only tools that we choose to take with us or not to take with us on the journey. When dealing with nature, we must be fluid and be prepared to pick what works best for us from the different models, and let them hybridise as they do in nature, naturally. Find a way to suit you.

Perhaps the best model of all to come to the fore in the latter years of the twentieth century has been Permaculture. A phrase coined by its founder Bill Mollison, permaculture (permanent agriculture) is 'the conscious design and maintenance of agriculturally productive ecosystems which have the diversity, stability and resilience of natural ecosystems. It is the harmonious integration of landscape and people providing their food, energy, shelter and other material and non-material needs in a sustainable way.'

I took my first permaculture course some twenty years ago, when the movement was very young in the UK. It was on a community in Wales where, for two weeks, a group of about fifteen of us learned the idealistic and forward thinking ways of this integrated system – a specific design system for living, if you like, and one that has much to commend it.

The basic tenets of permaculture are that we have to look after ourselves and learn to lessen our dependence on external inputs. This is absolutely key for the preservation of energy in a time when it is quite clearly running out.

As a complete system and a theory for sustainable living, permaculture covers design, building, farming and growing, energy, economics and just about everything you can think of in terms of ecological living. It is practised

Left: A turf roof blends into the landscape as well as insulating and cooling the house.

by millions of people all over the world, many of them unwittingly, such as tribal and indigenous communities who are free of outside influence.

An example of these communities would be the forest or homestead gardens found all over Asia and Africa but particularly in Bangladesh, Nepal, Kerala in south India and Sri Lanka. They work like this: the forest is the most stable environment on the planet. Here, animals, plants, soil and life of all kinds, great and small, come together for the mutual benefit of one another. They feed each other, provide shelter and co-exist in harmony.

Part of this system are the 'natural seven stories of the forest'. At the very bottom is the root zone. In south India, for example, this might be the tuberous root of the ginger plant. Then there is the ground cover layer, which might be lemon grass or vetiver. After that comes the small shrub, which could be a citrus plant such as an orange or grapefruit. Next is a small tree such as a pomegranate, after which comes a medium size tree such as a jack fruit, a silk cotton tree or an avocado. Up these are found the climbing plants, such as pepper or the areca nut vine. Finally, overseeing everything is the canopy layer, in the form perhaps of a giant rosewood tree.

This is how forests work with every species compatible with the climate and the nature in the vicinity. What man has done is to tailor this system to his needs, knowing full well that the order should still be there because it works naturally, the evidence is there for all to see. Forests are self-sustaining and the forest farmer knows this, so he goes about planting his own. What he gets is food, fuel, fibre, fodder for animals, shelter, medicine, timber for processing into everything from furniture to tool handles and the considerable peace of mind that he can provide for himself, his family and his animals.

I have worked within this system and seen it flourish to amazing effect, particularly in Tamil Nadu, south India. One of the reasons it works well is that the climate is such that plants grow quickly and for most of the year temperatures are consistent and life seems to go on. This is much harder for us in northern Europe, where we have quite distinct periods of slow growth when plants can go dormant for months at a time. That they need to do this to set fruit buds is neither here nor there; life appears to come to a standstill and there would appear to be a hiccough in the system. There isn't; there just appears to be one.

One of the most astonishing places where you see permaculture in action is Bangladesh. I bring this up as an example because Bangladesh is a country dominated by water. It can be subjected to appalling typhoons, which sweep up through the Bay of Bengal and the mouth of the Ganges and wreak havoc across the southern half of the country. The country is dominated by rivers, and travel is done by boat. The traditional rural dwelling, known as a bari, is raised up on a mound to keep the flood waters at bay. In the flood season, some of the baris become their own islands and often move with the river. On them are all the needs of the household in the form of trees, shrubs, arable plants and animals and all that it takes to feed them as well as the humans. The fact that this country, with a population of nearly 100 million and bedevilled as it is by the most extreme climate, can come up with a solution of this complexity is a testament to the Bangladeshi people. One would imagine that the drain on soil nutrients caused by the constant hammering of the torrential monsoon rain must be considerable, but the farmers and homesteaders know that the soil is protected by the canopies of the trees as it is in the forest. This method of protection is designed by nature herself. What the farmers have done is take their own template from what they see in nature. A more complete permacultural picture could not be imagined, but these are a people some 90 per cent of whom live in rural areas and are engaged in some kind of agriculture. We urgently need to reconnect with nature if we are to pick up the kind of skills that we will need if we are going to survive.

Taking a wholehearted permacultural approach may seem too much. It is important to work out what suits your lifestyle and what you can manage. I keep on stressing this, as do McFadden and Groh's in *Farms of Tomorrow Revisited*. One of their key pieces of advice is not to work too many hours: 'farming is labour, craft and art. The art arises out of a deeper understanding of nature based on thorough, ongoing observation, reflection and meditation on all surrounding natural phenomena and processes. If the farmer is working too long hours he lacks the leisure for this observation and he loses his art'.

The same is true of a sustainable life. The world's problems will not be solved in a day, and as so much of gardening is down to close attention to detail, try not to do too much. When you are dealing with nature, you need time to be in it, to watch and observe the subtle changes that take place each day. It is something not to struggle against and dominate but to cherish.

THE FIRST STEPS

So, with all these various models to work with and inspirational ideas, you should be keen to make a start. The idea is to become more self-reliant, to be more 'sufficient'. Every inch of space, front and back garden alike, can be used to grow food and rear animals. It is not a life of drudgery but incredibly good fun and opens up a world that most of us mortals never enter into.

The urgings of the biodynamists and the Fukuoka faithful to look closely at nature and see the wonders therein are not far-fetched but entirely sensible. The basic tenets of good gardening are centred around healthy, living soil – there is no other way. It makes no difference whether you cultivate one acre or ten thousand. The principles of nature remain the same: cause and effect.

Few people have summed up the concept of 'sufficient' as well as Robert Kourik in his book *Designing and Maintaining Your Edible Landscape Naturally*. He says, 'Roll back part of your lawn and renew the age old tradition of surrounding a home with a productive landscape. Using designs that suit your spare time, edible landscaping is a convenient way to grow vegetables, berries, fruits, nuts and ornamental plants in attractive and harmonious groupings, without the use of dangerous chemicals. As you nurture your edible landscape, it will sustain you and your family with benefits that go far beyond good food'.

It could not be better put. In one paragraph, he gives us permaculture, organics, ornamentals, fun, change, landscape, food and a wonderful home environment and also that key point 'design to suit'. This is where it must begin – with something that is manageable and enjoyable. If it becomes a drag, then the game is up. The ones who really know and have tried it always point this out because it is absolutely true. You want to have fun with your children in your garden and on your plot; if it is all work all the time, that will not be possible.

Above and Right: Design is the prerogative of the individual. Make it the first priority.

GROWING

GROWING YOUR OWN

Plants need air, light, food and water. If you supply them with these – and the optimum conditions in which to access these – you have every right to expect that they will grow. It sounds simplistic, but if you have taken the decision to grow plants for food – or indeed purely for aesthetic purposes – then the more basic this undertaking sounds, the more encouraged you might be. Virtually all living things need the same essential treatment, whether animal or vegetable. What is more, plants need love, attention and company.

The various issues that arise around the concept of 'sufficient' touch people differently. For some, the most pressing issue may be energy and its uses; for others, it might be waste. For me, the heart of the matter is food. How we think about it, procure it, consume it and – for the purposes of this chapter – how we grow it, store it, and make best use of what we grow. Yet this discussion is not just about food, it concerns social welfare, soil health, human health, the diversity of the planet, or plain love of nature; the connection with the earth and it inhabitants, which the American biologist and theorist E.O. Wilson called *Biophilia*. Food can be grown, smelled, touched, felt, listened to and tasted. It sustains us and our families, animals and the land, the area, the neighbourhood, our own holdings and the tiniest of spaces. The planting of a garden and the growing of food crops encourages wildlife and is a link in the great web of life that is essential for the balanced functioning of our planet. What is more, almost all of us can grow some food and make a huge difference to our lives and those of others around us.

Just as those who practice Permaculture (see pages 43–45) seek to mimic nature by design, so too a garden, a balcony or a smallholding where plants are grown can contribute to nature. By mimicking nature, working with and encouraging her, whilst at the same time producing food, we can explore and understand so much of the natural way of life. This is a long way down the 'sufficient' path. When you first harvest a ripe tomato – the smell of which you could not possibly imagine existed before – you will realise the truth of this.

One of the main tenets of 'sufficient' is that when we grow plants we ask only what they are capable of rather than forcing them to overproduce. It is the same with animals (but that will come later). With correct cultivation, plants will give willingly of themselves and we must learn to accept their bounty rather than trying to get more just because we can. This approach will lead to healthy and vigorous plants that will nourish us and our gardens, and it applies in the commercial sector just as it does at home. The overstretching of land and its natural resources within industrialised farming has led to the almost total breakdown of the system.

At first, the thought of 'growing your own' may sound like incredibly hard work but – as many have pointed out in the past and I will do so again – there are ways around this by means of clever design, the use of multi-functional techniques for growing and irrigation and, indeed, the use of plants that can offer more than one benefit. The process of growing your own should be fun and the journey towards achieving your goals as rewarding as the end result. Certainly, it takes application and trial and error, but each of us will find what works and what does not. This book aims to aid us in our cultivation, wherever or however we live.

ORGANIC VERSUS NON-ORGANIC

Most often, the word 'organic' is used in the context of food and its production. It has become a generic term for produce grown without the use of chemicals. 'Organic' is also a legal term of food certification, and various worldwide bodies, such as the Soil Association in the UK, grant licences to growers so they can label their food as such.

In the past, I have been a registered grower of organic food to certification standards, and my whole thinking about the way to cultivate crops is based around the perception that it should be done without the use of chemicals and in rhythm with nature. There is no place for undesired chemicals in the food chain; already quite enough pollution has been emitted through our water courses and into the atmosphere. (Some argue that it is already too late to grow uncontaminated organic food, as genetically modified organisms are present throughout our food chain and can never be removed because our bodies do not recognise them and they do not break down).

After the Second World War, during which food had been in short supply, an enormous range of artificial fertilisers and chemicals were introduced to deal with pests and diseases and improve food productivity. This new era of agrochemicals promised food for all but proved an unmitigated disaster, particularly in Asia, where poverty-stricken farmers were more than happy to try the heavily subsidised and cheap chemical products, in what became known, ironically, as the 'Green Revolution'. It soon became clear that these artificial fertilisers destroyed the soil life and therefore its fertility, and the pests soon developed resistance to the pesticides. The old traditions of crop rotation to build soil fertility and break the cycles of pests and disease, (in practice since farming began in the civilisations of the Tigris and Euphrates Deltas) were swept away, to the detriment of agricultural communities the world over. The Green Revolution had proved a disaster, but today, even with the huge increase in popularity of organic food, the organic farmer and grower remain in the minority and the agrochemical giants remain a powerful lobby.

In the twenty-first century, when we see land under intensive agricultural production, we must remember that its productivity is dependent on a barrage of chemical cocktails, from artificial fertilisers, pesticides, herbicides

Right: Velwell Orchard; a bio-dynamic holding amidst the rolling hills of South Devon, UK.

and fungicides, with little thought given to crop rotation or the addition of organic matter to improve soil fertlity. We are told by government scientists that these chemicals are safe and not harmful to human health. I don't believe it. I believe that if we want proper clean food, we should grow it ourselves by the organic method. That way, we know that the food we are eating is uncontaminated and arrives at our table fresh and in its most vital state.

The subject of genetically modified organisms (GMOs) in agriculture is one that arouses strong feelings both for and against. Since man began to farm the land, he has assisted the natural process of fertilisation to increase plant production. However, genetic modification facilitates genetic crossovers between species that cannot cross-fertilise in nature. This tampering with nature gives cause for grave concern, as does the insertion of genes from one plant to another to achieve a particular aim, such as better storage and longer shelf-life. The pro-GMO lobby will cite the increased ability of these modified crops to feed the hungry and reduce the level of pesticides required to grow them as justification for their creation. Nonetheless, this is a frightening development. As usual, the motivation behind the introduction of GMO crops seems to be the greed of the manufacturers and distributors of such products to the detriment of the land and the people.

This section of the book will concentrate on growing crops by the organic method, simply because that is the way I prefer to work and have always done so. There is little point trying to justify the superior quality of organic foods through facts and figures. It is an inherent feeling and a belief that correct cultivation – through the building of soil fertility and the recycling of nutrients around the farm or garden – leads to healthy plants, which will inevitably lead to healthy people. Although taste and flavour of food often has a great deal to do with freshness, I am, nonetheless, an organic grower and would encourage others to adopt this approach in whatever way is possible.

The food scares that have troubled Europe and North America in recent years have played into the organic producer's hands and made their case stronger. If you feed bits of cow to other cows, it does not take a genius to conclude that something might go wrong. Similarly, if you fail to recognise the soil as a living organism that needs feeding – as all living organisms do – then expecting the soil to nourish plants without being fed is a hard task.

Yet this is what has happened over the decades, particularly since the Second World War. The soil has become merely a 'rooting medium' for plants rather than a resource that nourishes them. The plants themselves are fed by artificial food in the form of expensive synthetic fertilisers, which are produced in energy-inefficient factories by means that use enormous quantities of fossil fuels. The dust bowls of the American wheat-growing belts are testament to this. Soil erosion has become chronic because the soil has no life. Soil without life turns to dust.

All the nutrients that feed the plant are drawn into it from the soil via the roots, and in combination with light, chlorophyll and air, the plant is allowed to photosynthesise. Such is the importance of the soil, its health and care, that the whole crux of the organic argument hangs on it. It is no coincidence that the UK's premier organic certification body is called the Soil Association.

All this said, one must make sensible ethical choices when opting for organic products or deciding to eat only those which have 'clocked up' few food miles. For example, gardeners are discouraged from using peat-based composts because peat is a finite resource and is considered to be unsustainable. Yet when making our decision, we rarely stop to consider the Irish peat farmer and how he will cope with the closure of his industry. Similar are the human ethical issues surrounding food miles: why should we ignore the tiny African farmer who is struggling to export his crops to a global market, while we drive to buy an English organic product that may have consumed as much if not more energy in its processing, packaging and transport?

The principles of 'sufficient' are based around fairness and equality, both in the way we grow our food and the way we live our lives. Overproduction has no place in this system; organic production favours life over greed. We have to relearn the skills of recycling nutrients rather than relying on external supplements to help us grow our plants. We must return to being resourceful and learning how we can make the best use of what we have without having to take it from elsewhere. This can be achieved by simple means, starting at home, by converting an area of lawn into a vegetable patch, or building a worm composting box to recycle our kitchen waste rather than throwing vegetable and fruit peelings away.

If we choose the organic way, we set out to follow the basic principles of 'sufficient'. We have chosen to become more self-reliant by using our available resources and by learning how to turn these simple resources into plant food. It is an integrated method of cultivation, and one that was in place long before any chemicals to aid cultivation existed.

SITE AND DESIGN

Let us assume we are starting from scratch. Before you can begin preparing the soil and growing vegetables, you need to get to know your site, familiarise yourself with how it functions, and discover how you may be best equipped to work with it. The design and layout of the growing spaces becomes more obvious when you know what it is that you can and want to grow on your site – it is the plants that take up the most space and so ultimately they shape the design. Clearly there are elements to a 'sufficient-led' life that do not involve plants and that need space in the garden, but it is predominately plants and their cultivation that requires design.

The first external influence to take into account is the weather. How does it behave where you live and what is the characteristic climate of the region? Obviously, some crops grow better than others in certain areas – for example, apple trees in Kent – but for general purposes you need to know about the rainfall, the wind and the sun. Do cloudy and moisture-laden weather patterns dominate, or is it dry and bright for long periods of time? This information is important for helping you to decide what vegetable crops you want to grow and how you lay out the land to best advantage.

What happens in the winter is vital too. What are the incidences of frost and snow? To help determine planting times, you are going to need to know when the first frosts can be expected in autumn, and when the last frosts are behind you at the start of the year. Is your site in a frost pocket and, if so, where does the frost drain off too? You need to know how these kinds of weather patterns affect not only your area, in general terms, but the chosen site of the vegetable growing plot itself. What is the aspect of the plot in relation to the sun? Where does the sun come up in the morning and go down in the evening, and how much of the site is shaded by trees or buildings, and for how long? What protection is there from the wind and what direction does the prevailing wind come from?

Successful growing owes a lot to observation and attention to detail. Look around your site and look around your neighbourhood. See what is growing in other people's gardens and be prepared to ask questions. Local knowledge is very valuable. Gardeners with productive vegetable gardens are a dying breed, but a generous lot, and wisdom from the older generation – who may have been gardening for a lifetime – will aid you and probably make you some friends in the process. The concept of 'sufficient' is an inclusive and interactive one, and sharing resources and information is a prerequisite.

Only once you have a firm understanding of the conditions in, on and around your site can you start to think about the best layout for whatever it is that you want to grow. You must also think carefully about how you will move around the site as you carry out tasks. Will your paths need to be the width of a wheelbarrow or will they need to be wider if you are using a vehicle? How will you get large or heavy materials, such as manure on and off the site, and how will you move them around the site itself, particularly when it is wet? Will you be keeping livestock and how will they fit in?

Next comes the understanding that you may have to make considerable changes to the site as it exists in order to transform it into a productive site for growing vegtables. There may be tall trees that are casting too much shade over the site and need some branches removing to increase light levels, or you may need to plant more trees to add more shade or to act as a windbreak. It is very important to come from the realisation that you get only one chance to plant a tree. Once trees are planted, they are in and should be left alone. They are not moveable structures like polythene tunnels or sheds that can be dismantled and put up in a different location if you change your mind.

Any building removal or repair of structures, any land drainage and any hard landscaping should be tackled at the same time as services are put in, such as water for irrigation and electricity for greenhouses, sheds and polytunnels.

Water on a site that is used for productive gardening is particularly important, especially if you are using a greenhouse or a polytunnel to boost crop production and extend the growing season. The services (water, drainage, power) have to go in first and whilst the use of water is something that I will talk of later (see pages 250–56), there is no denying a mains water supply is essential. It is no good just relying on water butts filled with rainwater.

Whatever you do with your site, new or old, make a plan. Even if you are unsure exactly what you want to grow, it helps to know what it is you are aiming to achieve. Your dreams and goals will drive you and, while sometimes they may not be attainable, the path towards them will be made so much easier if you have a plan to follow.

GARDENS

You may dream of creating a forest garden and being able to harvest hardwood timber, or you may simply want a few raised beds with Khaki Campbell ducks free ranging and hoovering up slugs, but a garden needs a vision, just as a site needs overall planning. So before you get started, it is worth spending time working out what you want the place to look like and how you are going to work it. The final design will come when you know what you want to grow, and this will put the finish on the job. Nonetheless, some key decisions have to be taken early so, for example, if you really don't want to use a lawn mower or strimmer on grass paths – as I have decided – then you will have to use some sort of mulch or hard landscaping for paths instead. That entails working out where that mulch is going to come from and how it is going to be laid.

One of the most important considerations, which runs in tandem to layout, is to try and decide which crops you want to grow and how much you want to grow of each crop. That way you can make a well-informed order from a seed company and get started, while also getting some idea of the design you will need to accommodate the crops.

Every part of our house and garden that has access to light is potentially a growing space and, once we start looking in this way, we can see that the opportunity to be productive is enormous. Every drainpipe on a sunny southern wall could support a kiwi fruit and every empty container could hold a couple of parsley plants. What is absolutely for certain is that if you are a first-time gardener and you get bitten by the bug you will have plants in every available free space, both inside and outside. When you fall into gardening, you fall hard, so be prepared.

This 'passion for production' makes design doubly important because space must be used wisely. From a design point of view, the plants that we choose need to work hard and serve a dual purpose, particularly if space is limited. For example, a plant that gives us good scent in winter but lacks appeal during the rest of the year, can justify its place in the design by acting as the support for an annual climber, such as beans, during the summer months. Likewise, while a roof provides shelter, it can also be used as a growing space for plants to attract wildlife and cool down the house in hot weather. We must also learn to be ruthless and make sacrifices for the sake of design. We may want only a dozen lettuce plants, but we stubbornly hang on

'Zoning' is a means by which we can design a plot to be energy efficient. It is very simple in its application because it works on the basis that the parts of the garden you visit most often should be closer to where you spend most of the time – and that place is the house.

So, when you want a handful of chives on a wet, dark night in winter, you do not want to have to stumble the entire length of the garden to find them. The 'frequent-visit' parts of the garden – i.e. the intensive vegetable beds – should be close to the kitchen. There is nothing more miserable than a long walk in the wet and the dark when the torch has run out of batteries, to dig a cold, mud-caked, canker-ridden parsnip. Have it close to the house instead.

Left: The abundance of a natural garden.

to 50 until the rabbits would not even look at them, citing the need for spares, extra for barter and so on. Lose them – they make good compost and the garden will look all the better for it.

Remember that you need an area set aside for the propagation of plants. Annuals are raised from seed, and whilst many can be sown directly in the garden in an open ground seed bed, the more tender seedlings that are started earlier in the growing season, such as cucumbers and tomatoes, need indoor protection from frost. They should be raised under cover, whether in the house, in a polythene tunnel (not always frost-proof) or under glass.

The truth of the matter is that unless you have a very large garden or some further land, it is very likely that you will run short of space, so you need to use what you have wisely. And you should use it. I am of the persuasion that says the front garden should come under vegetable and fruit production and that productive gardens should not just be placed at the back of the house. What shame is there in having beautiful climbing beans growing in the front garden where once there were roses? You can keep some roses, of course, and they will happily grow and climb up poles with the beans.

ALLOTMENTS

For those of us without a garden but with a desire to grow fruit and vegetables, there is much to be said in favour of leasing an allotment. This is a piece of ground usually of a size in the region of $\frac{1}{10}$ acre (0.04 hectares), which is in the ownership of the local authority. The idea of making land available to people for food production sprang from food shortages through history. The earliest allotment schemes were started in the UK in the eighteenth century. Germany soon followed, and today allotments (the English name for such land) can be found all over the world.

In the US, there is a strong tradition of community gardens. This is a similar system to that in the UK and Germany, based on the leasing of a plot of land to individuals by public bodies. Often sales of produce are restricted by the terms of the lease (and the small print may deny the allotment holder use of the ground for leisure purposes), but this should not deter the 'sufficient' grower. Allotments represent an opportunity to gain access and use of some land without having to purchase it, and start cultivating.

Right: There is always work to be done; know your 'sufficient' limits and requirements.

In the UK, the popularity of allotments is increasing to the point of waiting lists for potential applicants, but it is worth the wait because these are gardens that have often been well tended over many years and if you are lucky enough to be granted one, the chances are you will find the soil in good heart.

The sharing of information and resources is common amidst the allotment community, and the value of this collective knowledge cannot be underestimated. Tapping into these years of experience is a great help, especially for those who are just starting out. It is not often we get the chance to wander freely through numerous gardens and engage with those who care for them. Here, you will pick up the local information on just about everything, from soil type to weather conditions, which is so valuable to the success of a productive garden. Seed exchanges, loan of tools, materials and labour are also generously offered by members of allotment communities.

If your application to the authority for an allotment is unsuccesful, the next step is to visit your nearest allotment site and discover whether there are any plots that are uncultivated or abandoned. Armed with this knowledge, you may be able to discover who the leaseholder is and whether they would consider giving up their space or, indeed, allow you to take over a portion of it. Do not be shy. You have to be resourceful and find the help yourself.

In the US and Japan and increasingly in the UK, the idea of 'community-supported agriculture' is becoming more popular. This is a system whereby the individual has a stake in the farm and receives produce in return. It is a good way of becoming involved in growing, understanding the provenance of the food you eat, and seeing food production on a large scale. It also allows the individual to have a say in what is grown, unlike a box scheme from an organic farm where you have to take what is on offer, like it or not.

GROWING IN A SMALL SPACE

The smallest spaces open up opportunities for us to be very resourceful in our attempts towards a life of 'sufficiency'. Throughout the book, I will indicate what are suitable crops and cultivating techniques for small areas (as well as irrigation and feeding systems), but to begin with you must not be downhearted just because you do not have a large area in which to practice. You might be surprised by just how productive you can be in a limited area.

Right: Never underestimate the smallest space.

For those with little open ground, growing plants in containers can be a useful solution. However, not all crops are suited to the restrictions imposed by this method of cultivation. Container gardening is not as straightforward as gardening in soil in open ground because plant size and support and nutrient availability all need careful consideration. We have to learn the appropriate skills, but with some attention to detail this is very rewarding. An added bonus is the fact that the plants in their pots can be moved to different locations to benefit from the best environmental conditions.

For those with only window sills and, at best, a window box as a container, the difficulties become more extreme, but remember that for those who are propagating plants for a garden and have no outside means of doing this, such as a polythene tunnel, a windowsill is the only answer.

Perhaps you will have to subscribe to a box scheme for your fruit and vegetables. Even so, a fresh herb plant, like thyme, transplanted and grown in a window box can supplement and enhance your cooking and is worth the effort. As much as anything, it is the concept of being part of a movement that wishes to embrace 'sufficiency'. The more we think about change and actively engage in it, the more it will happen.

Perhaps if you live in a flat in a block you might be in a position to start the process of change. I worked for a while for a landscaping company in London which had the maintenance contract for the lowly gardens attached to a block of high-rise flats. It was a rough, poverty stricken area with many of the problems associated with inner city housing. Yet at the foot of one of the blocks a group of Asian families had negotiated with the council to remove some existing plants and start growing food crops for themselves. Tended by the family members, these gardens were productive year-round and proved to be immensely popular with other locals.

Today there exist a multitude of ways by which we can attain land for production. In times like ours, when we can no longer rely on others to provide for us, we have to take opportunities and stand up for ourselves. If this means negotiating with local authorities then remember that now, like never before, are they prepared to listen. In my home town, there are no allotments and people are desperate for land. Petitioning the local authorities has led to their capitulation or at least their understanding that this is a problem that needs addressing. The result is that a land search for allotment sites is now under way. We do, after all, live in a democracy.

TYPES OF CULTIVATION

This may seem an odd heading, but there are many ways to grow many different types of plants, and different disciplines for each. I have talked already of my preference for the organic method of cultivation, but there are variations on that theme too.

One of them is the 'bio-dynamic method', which is based on lunar cycles in combination with the vital forces and energies of certain herbs and plants used to invigorate land and compost. Taken from the teachings of the Austrian anthroposophist Rudolf Steiner (1837–1913), this system is the subject of growing enthusiasm amongst farmers and growers because it is perhaps more comprehensive and in tune with nature and her rhythms than a straightforward organic approach.

Permaculture encourages emphasis on perennial or permanent crops, such as trees and fruit bushes, with the view that this is more energy-efficient and sustainable than planting annual crops, with less disturbance of the soil. Food from trees, such as fruit and nuts, and perennial vegetables are relied upon for production, while plants are used to provide materials for building, fodder, medicines and fibre. Mulches are used continually and plants are encouraged to self-seed. A more natural way of gardening or growing is the aim.

This leads on to 'forest gardening', in which gardens are planted to mimic the seven natural layers of the forest. Food is found on every layer and the harvest includes everything on offer from the trees, shrubs, climbers, lower level plants and the root zone. This model is taken from the home gardens of Asia and Africa, where tiny plots of land mimic the natural forest and are highly productive.

My own cultivation model is a hybrid of all of these, with a few techniques of my own added in. Personally, I believe that the principles of Permaculture as far as growing food is concerned are better suited to warmer climates. In the temperate zones of northern Europe and North America, we struggle to find perennial crops that nourish us through the darker months, no matter how many ingenious food storage tricks are employed. I want to grow as much as I can in the way of annual crops (and make use of protected cropping to do it), with the aim of stretching the growing season from an earlier spring and into a later autumn via good use of polytunnels, greenhouses and cold frames.

For those who have one, the sun room or the conservatory also present excellent growing opportunities and an ability to extend the seasons.

What is most important of all is that we are nourished by what we grow and by the process of the task. That we learn skills as we go along and use them to decide what is best for us is the benefit of having all these cultivation models to work from. A key part of 'sufficient' is that we feel we are equipping ourselves to be more self-reliant, and one of the best ways to do this is to watch and learn from all those different disciplines.

There is no right or wrong way on the path to 'sufficient', just a desire to care for the earth and allow it to produce freely for ourselves and those around us.

DIG VERSUS NO DIG

Let me be clear that this curious phrase 'dig versus no dig' does not have any relevance during the set-up phase of a new growing project. Civil engineering of a kind will almost inevitably have to happen in the early stages to enable you to get the type of garden and growing beds that you want. By this, I mean you might have to move earth, plants and even buildings, and cause some upheaval in the process, to end up with the right garden for you.

'Dig versus no dig' refers to the manner in which you approach the cultivation of the soil. Once you have the system that appeals to you in place, whether it is a vegetable garden of raised beds or good old-fashioned borders, there is an argument that suggests there is no further need to disturb the soil. That everything you grow in the garden, edible and non-edible, can grow through a permanent cover of mulch. This is, after all, what happens on the forest floor, one of the most stable environments on the planet.

Our aim should be to have soil that is alive, and it follows that constant tinkering with it by way of annual digging will upset that life. However, as much as we need to protect soil, it is clever stuff and has ways of protecting itself. Soil will form its own 'skin' if left bare. A compost heap is the same. If you go digging around in a compost heap, you will find that the top few inches are usually made up of a tough, dry crust and underneath lies the damp, brown gold. If you create a strong and fertile soil, you will find that even in the hottest of summers, when the top inch or two may seem dry and crusty, the soil beneath the dry layer will remain happy, moist, and teeming with life.

There are few places where new soil is permanently being created. One such place is under still water where silt gathers naturally. A second is under permanent pasture and lawns where the continuous cycles of life and death of the various species of grasses and plants help create fertility and therefore soil itself. And the third is under a mulch as on the forest floor or on vegetable and flower beds in the garden. The presence of the mulch itself, assuming that it is of an organic nature and so can break down, will be constantly creating soil while as a surface cover it is protecting what is going on beneath and allowing it to happen at a favourable rate.

When it comes to sowing annual crops – especially small-seeded ones that need to be sown directly into the soil – it is difficult to garden where there is a heavy mulch of a material, such as straw. Therefore each year before planting, a little raking out of a seed bed is necessary, but that is probably all. Most perennial crops, and particularly fruit bushes and trees, are well served by a mulch, and the soil around them will remain undisturbed. In my opinion, the less digging you can do the better, for your soil, your tools and your back.

What really matters is that you find the way that suits you, a balance. The world of growing and gardening is wracked with contradictions and snobbishness regarding the way we should work and what everything should look like, ignoring consistently that it should be a process of fun and enlightenment. Do what is best for you and be happy with that.

TOOLS AND EQUIPMENT

Tom's Cornish Shovel

Where I come from, we use a long-handled, point-ended shovel designed for digging holes, trenches and paths and moving large quantities of soil. It is a tool used all over the Celtic areas of the world, in Cornwall, Devon, Wales, Brittany in northern France and, to some extent, the US. It has been in existence for centuries for the very simple reason that if used correctly it takes the strain off the back. The effort in heaving a shovel full of anything heavy is taken by the shoulders, the stomach muscles, the arms and the thighs. In other words, it is a safe tool to use and one that, once you get used to it, will become your spade of choice. The idea of a short digging spade is a modern invention, as is that of the fork. In the past, both the fork and spade had long handles.

Right: The Cornish shovel stands alone for me as the tool of choice.

It sounds so obvious that we are going to need tools, but which ones? Let us start with the biggest. Unless we have a sizeable holding and require a tractor and trailer to haul materials around the site, the other large cultivation tool driven by a two-stroke engine that many smallholders consider is a rotavator. Let me be quite clear about this – a greater curse was never bestowed upon the land than the whirling tines of a rotavator. It is a beast whose blades turns over and destroy exactly the soil life that we, as gardeners, are trying to build and protect.

The matter of the soil, its care, protection and enrichment is dealt with on pages 78–85, but from the outset it is critical to appreciate that the soil is the medium by which everything we grow depends upon. On that basis alone, a machine that is an enemy of the soil cannot be tolerated. A rotavator may render the soil light and fluffy in texture, but the damage done to the soil structure is disastrous.

On the same subject of protecting the soil, and still with tools, we have to look at 'tools' that protect the soil in the growing beds. Walking on freshly cultivated soil and crouching down to plant, prune, or carry out any task will compact and damage the soil. It may sound like neurosis, but just try digging out a planting hole on a spot where you have been standing for ten minutes. It takes some doing. If you intend to walk on the soil, the most important 'tools' to have at hand are walking boards. If you spread the load by placing walking boards across the soil, you will cause little or no soil compaction and water will not gather in your footprints and create mud. The only repair needed is to rake out the marks of the boards left on the soil. The boards can be anything from old bits of skirting boards to scaffolding planks, but remember that the wood can get heavy after rain and they must be lifted off the beds when you have finished working because they make ideal hiding places for slugs.

Machinery and walking boards aside, it is also worth looking at some of the more useful hand tools, as most of the gardening under discussion here is on a relatively small scale. Clearly, a spade or a shovel and a fork for loosening up and perhaps occasionally turning the soil are essential, but there should be no digging just for the sake of it. The precious worms and micro-organisms that are busy under the soil surface will only be exposed and taken away from their valuable work by unnecessary digging. I make no apology for referring once again to my 'no-dig' philosophy.

A stiff, straight-ended landscape rake is very useful. This is the tool that renders the soil to a fine tilth for seed-sowing and planting and grades the soil to the level required. The rake is an excellent all-purpose tool that the gardener can not manage without. A distinction must be made here between a landscape rake and a spring rake of the type with bendy tines, used most often for light work, such as raking up leaves. One to avoid for use on the soil is the wooden hay rake, which, while able to move large quantities of soil at a time, has short wooden tines like pegs, which have a habit of breaking easily.

There are two specific types of hoe that I recommend, both of which have their uses beyond the removal of weeds. A draw hoe has a long-handled shaft with a downward pointing flat blade on the end, the action of which is to lift weeds and draw them towards you. It is not an easy tool to use, as it tends to move more soil than is necessary. What's more, the weeds that have been disturbed can stay hidden in the soil, thereby giving them the chance to re-root. However, it is very useful for making seed drills or shallow trenches, as it is easier and more natural for the operator to pull a hoe across and through the soil rather than to push away, as when using a push hoe.

The key to success with a push hoe is to make sure it is kept sharp so that it can cut the weeds which are left lying on the surface. Keep a rough sharpening stone to hand to sharpen the cutting edge of the hoe as you work. Sometimes the hoeing action will lift the weed out of the soil, roots and all. This is why hoeing should be done on hot sunny days so that the exposed roots of the weed will shrivel up in the heat and the weed will die. On damp days, the weed might re-root and you will have failed to achieve anything.

The smaller hand tools, such as a sturdy trowel and a hand fork are also invaluable, but to ensure a long and happy working relationship, the tools should be made by a well-established manufacturer. Avoid hand tools that look gimmicky and as though they have been designed for a specific purpose. These will inevitably be of a poor design and will be of little use.

It is worth searching secondhand shops for old gardening tools. Many of these old tools are no longer manufactured, largely because we do not garden with the intensity that previous generations did. I was delighted to find a hand-held short rake, about 60cm (24in) long with strong tines, in such a place. This hand-held rake can be very useful for reaching beyond arm's length when you are hard at work on your knees or for reaching over plants in areas where you do not want to tread. It gives that extra bit of leverage.

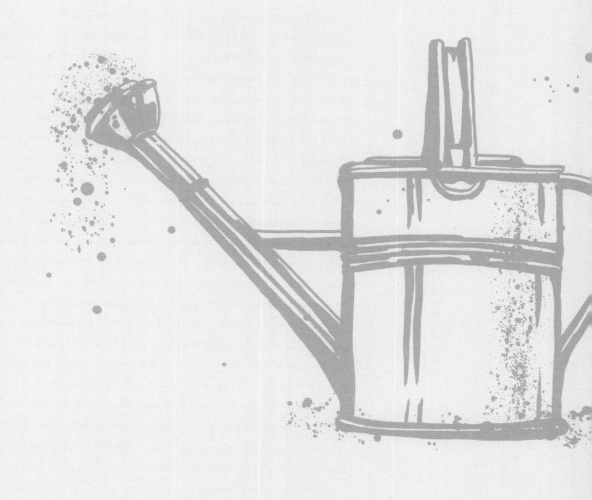

'A good watering can with a fine rose is vital for watering young plants and delicate seedlings and, here again, it is worth investing in a quality brand with a brass rose.'

Cutting implements are essential companions and the list is headed by a reliable pair of secateurs. 'Pencil-thickness' wood is the biggest size recommended by manufacturers for cutting with secateurs. Top brands are expensive, but this is an investment and good secateurs are worth looking after and servicing regularly. That way, they will last a lifetime. A sturdy pair of scissors is almost as valuable, though they should not be subjected to the sort of work better suited to secateurs. A small folding pruning saw is a very useful multi-functional tool, while a pocket knife is essential and should be kept sharp. As students at horticultural college, we were dismissed from practical sessions if we failed to attend without a pair of secateurs and a pocket knife. It is pointless heading into the garden with the intention of working without at least one or the other; there is always something to cut and bring indoors.

A good watering can with a fine rose is vital for watering young plants and delicate seedlings and, here again, it is worth investing in a quality brand with a brass rose. The plastic watering can with a brass rose fitted into a plastic rim will not last. Also in constant use are bamboo canes for supporting plant stems and jute string for tying in shoots. All kinds of household containers should be saved for garden use. Glass jars are useful for storing seed and keeping them dry, while household waste, such as silver paper, makes effective bird-scaring devices. Plastic bottles can have a second life as mini-greenhouses or plant protectors. An invaluable tool is a hand-spraying bottle filled with water to mist plants and seedlings, or filled with ecological soap to control pests like aphids. A metal pail will also prove useful for carrying water and materials.

For propagation purposes, there exists a specialist set of garden tools to carry out the delicate work of growing and transplanting seedlings and soft-wood and hardwood cuttings. Absurd as it may sound, a dibber is perhaps the friendliest of all garden tools. I was distraught when I lost my hand-turned dibber made from a piece of holly wood. (The purpose of this small pointed implement is to dib a hole in compost to receive a seedling during the 'pricking out' process. Its action creates the perfect size and shape of hole for planting the seedling.) My dibber was a tool that I had become so familiar with that my 'pricking-out' technique suffered when I was faced with having to use a replacement. There are large dibbers too, used to plant vegetables such as leeks, which need deep, pencil-thin holes in the soil to keep the stems in the dark and blanch them. These large dibbers are often shod in steel or brass to make them smooth and easy to extract from the soil.

Also on the subject of propagation, it is worth making or buying wooden seed trays in which to grow the seed. The use of ugly plastic seed trays should be discouraged even though sturdy plastic trays and pots can be used over and over again. Biodegradable pots are widely available, but wooden seed trays are best – not only because of strength and rigidity – but also because they look pleasing. The aesthetics of 'sufficient' in the garden are important to me; it is not all about rubber tyres and black plastic just because they are practical. A supply of hessian sacks rather than paper sacks for storing potatoes is also a good idea. This beautiful material offers a good solution, as it lets the potatoes breath while keeping them in the dark and preventing them from turning green. Hessian is also a useful material in which to wrap large tender plants such as olive trees, to protect them from frost and for transporting materials to the compost heap. Again, hessian sacks look great, and this material makes good aprons, hats and coats if need be. When they are not in use, hessian sacks need to be kept dry or they go mouldy, and they must be kept away from rodents, who will otherwise set up home in them and eat them.

SEED COMPANIES

F1 HYBRID SEEDS

It is sad from an organic grower's point of view to see more and more F1 hybrid seed being offered organically. These are seeds from plants that have been crossed as many as ten times to achieve the required characteristics. They were developed to suit largescale production methods because elements, such as uniformity of size, time of harvest and shelf life are beneficial for the largescale grower, retailer and consumer. For the 'sufficient' grower what is important is flavour, dependability and the ability to save a seed from a plant for the following year. With F1 seeds, this will not happen. They have been bred to the point where their seed may fail to germinate the following season. Brussels sprouts are a classic example of a plant that has suffered at the hands of seed companies who specialise in F1 hybridising. Choose the Victorian open-pollinated varieties of Brussels sprouts and ignore the F1 hybrids.

Placing seed orders and choosing and using seed companies bears some discussion. Looking back on your seed growing successes and failures in the garden over the past year and planning which seed varieties you wish to try the following year, is always an enjoyable way to spend a cold, dark winter's evening. For me, it is filled with excitement and expectation for what the year ahead has to offer. However, choosing and ordering seed can also be a time-consuming and even frustrating experience if not planned well in advance of the growing season. There are many suppliers to choose from, and some seed companies will offer a better service than others in their ability to source the seeds, process and deliver an order.

You can pore over a stack of different catalogues to try and save a penny or two by comparing prices, but it is not really worth the effort involved – things will even out in the end. To avoid disappointment, it is time well spent to get your seed order in – however large or small – as early as possible before the growing season starts.

It is also certain that you will want to supplement your order at different times of the year, so plan ahead and have second choices in mind. To that extent, try and make friends with a seed company that has experience of growing what they sell. This can be very helpful in that they will know a good alternative if your chosen variety is unavailable, and will be able to advise on propagation and growing conditions. A little advice can go a long way.

Taking care of the seeds when they arrive is important. Your raw material is expensive and is hopefully the beginning of your own 'seed-saving' foundation – the idea being that you can collect and dry your own seed and bring about a considerable reduction in next year's seed purchasing bill. So treat the expensive package well. Small seeds that arrive in sealed foil packets store well in a cool dry place, out of direct sunlight, until needed. On delivery, seed potatoes need 'chitting' (sprouting before planting). Unpack the potatoes and set them out so that they each stand up on one end, in a light but frost-free place. On arrival, onion sets should be unpacked and spread out on a tray to avoid the spread of any fungus from rotting bulbs. Keep the onion sets out of sunlight. The treatment of seeds is largely common sense, but it pays to remember that they are light-, temperature- and moisture-sensitive, so treat them gently to avoid untimely and unwanted germination.

'Collect and dry your own seed and bring
about a considerable reduction in next year's
seed purchasing bill.'

CROP ROTATION

The cornerstone of organic cultivation, the rotation of crops around land has been in place since farming began. It works on the basis of the movement of annual crops from one place to another each year, and this achieves two things: firstly, it breaks the life cycle of pests and diseases; and secondly, it builds soil fertility. Two more important functions for any grower are hard to imagine. It is still in place today on conventional as well as organic farms because it works. If you were to apply for organic status or you intended to sell crops labelled 'organic', you would be required by The Soil Association to supply a rotational plan to indicate that you have a grasp of these matters.

When we look at crop rotation in a little more detail, the two factors of pest and disease control and fertility work together. Different families of plants are affected by different pests and diseases as well as having different fertilisation requirements. By putting plants in a rotational order, you can achieve both aims.

For example, a standard five-course rotation might involve five different plant families: *solonaceae* (potatoes), *brassicaceae* (cabbage tribe), *apiaceae* (root crops such as carrots and parsnips) *leguminosae* (peas and beans) and *asteraceae* (lettuce and other salad crops). All have different nutritional requirements and are troubled by diverse pest and disease problems.

Imagine for a moment that you had those five crops growing over the course of a season in four different beds. How would you organise them in a system of rotation that can work over a four-year cycle? Bed One holds potatoes, which require considerable feeding with bulky organic matter. They grow through the course of the season and as they are harvested they are replaced by the winter brassicas, such as kales, cabbages and sprouting broccolis. What happens is that the potatoes will have taken up the majority of the nutrients that were introduced by the organic matter (manure or compost), leaving behind enough for the brassicas but not too much so that the Brussels sprouts get too leafy and burst (blow) or young growth is troubled by aphids on account of sappy nitrogen rich leaf growth. Any pests such as nematodes (eelworms) or fungal diseases such as blight (*Phytophtora infestans*) find their food source removed when the potatoes are replaced by the brassicas and therefore cannot prosper and spread. Two crops in two different families have been grown in the same ground in one season.

The second season sees the introduction of the root crops, such as carrots and parsnips that are sown directly in the ground on removal of any remaining brassicas. These root crops fall in the *apiaceae* family. They are not desirous of rich soil. The roots wish to go straight down unhindered by too much in the way of nutrient, which is why they grow so well in relatively poor sandy soils. The nutrients that were introduced into this plot the previous year have been first taken up by the potatoes and then the brassicas, leaving it in a perfect state for the root crops.

By the end of the root rotation in the second season, the soil is in further need of enrichment, particularly by way of nitrogen. Fortunately there are members of the *leguminosae* family which can return this to the soil, most of which we find very palatable, such as early peas and French beans. They need no applications of compost at planting because they are able to feed themselves. This they do by taking atmospheric nitrogen and holding it in bacteria stored in nodules on their roots. It is released for their own benefit and for other plants around them as required. In the process, it saves the farmer or gardener major effort, and considering that nitrogen is one of three key elements for plant growth, the 'fixing' of nitrogen by plants in the *leguminosae* family has to be considered one of the greatest miracles of nature. The pest and disease problems that trouble the umbelliferous plants, such as carrot fly, find their food source removed when the peas and beans are introduced and so die out. The salad crops that follow the beans in the fourth year pick up the available nitrogen left by the legumes and use it up before the cycle starts again with potatoes and the manure that is applied to aid them.

Thus we can see that a four-year cycle involving five plant families growing in four beds will increase soil fertility and keep down pest and disease problems.

It is very comforting to know that there is a built-in natural self-protection mechanism that carries out these two essential operations in one go. It is as though it has been designed specifically for smallscale intensive production.

THE SOIL

MULCH

Like many areas of organic or
'sufficient' gardening, mulch
evokes strong feelings, particularly
because some of the materials
used for this purpose don't look
very aesthetically pleasing. The
needs of the soil are the issue, and
there are some suitable materials
available for mulch that do not,
I think, detract from the look
of the garden, as long as they are
of an organic nature. The most
important consideration when
choosing a mulch is whether or
not it is suitable for your soil and
your plants in terms of how it
breaks down. Mulches are broken
down by the soil life at different
rates, depending on what
materials make up the bulk of
the mulch. Carbon-rich mulches,
such as wood chippings, are not
ideal for growing beds, as they
rob the soil of nitrogen in the
effort required by the soil life to
break down the wood. But lawn
mowings (an easy mulch to
get hold of from neighbours,
councils, etc) are nitrogen-rich
and so will not rob the soil
because it is able to use its own
nitrogen to break down. Stable or
farmyard manure is made up ▶

The importance of healthy soil in 'sufficient' terms cannot be stressed enough. This is the medium in which we grow our crops and its job is to sustain them. Before we can discover what sort of soils we have in our gardens and what their requirements are, it is useful to consider what soil is and where it comes from.

The soils we have in our gardens started life as rock thousands of years ago. The process of weathering through rain, snow, ice and atmospheric gases broke the rock down, and in combination with the arrival of organic matter (and no one is really sure how it did arrive) soil began to form. Glacial melt still forms new soils today and if you were to dredge a lake in a valley of the Swiss Alps or the Appalachian Mountains, there would be soil, in the form of silt, gathered by run-off at the bottom. Ultimately, all soil is on its way back to the sea via our water courses, where it will form rock again.

If you dig down in your garden soil, the chances are you will find three distinct layers: topsoil, subsoil and finally bedrock. All three are in different stages of decomposition and therefore continually forming new soil. From a gardener's perspective, it is in the top layer – the topsoil – where the action is and where plants get most of their nutrition. As the landscape changed over time to incorporate farms, towns and latterly gardens, soil has been a constant in the change, finding itself part of different enclosures. Your soil will be peculiar to you, but its make-up will be largely governed by the geology of the area in which you live. The exception to this is situations where topsoil has recently been brought onto your site from elsewhere. If your house is a new-build, or you have just moved to a new property where the garden has been re-landscaped, try to discover if any of the topsoil has been brought in from another site and, if so, where it has come from and what it is made up of.

No matter what sort of soil you have, it is very important to discover whether it is acid or alkaline. (The acidity or alkalinity is known as the pH value.) Plants are pH-sensitive and, as it happens, most favour slightly acid soil, but a whisker either side of neutral will do for most plants. It is more crucial to know if you are gardening with extremes of either acid or alkaline soil so that you can ameliorate the soil accordingly.

Most soils, regardless of country or place, are made up of clay, sand or loam. There are exceptions, such as peat and adobe clay, but these are relatively rare.

pH is easily measured by a pH testing kit, but you will get just as good an idea by looking around your area and seeing what sort of plants are growing – whether acid- or alkaline-loving species. Local knowledge is very valuable and you will quickly glean from neighbours, the local plant centre or public garden staff, whether the local soils are acid or alkaline.

Whatever soil you have, and all soils are different, it is certain to need enriching. So what is it that actually needs enriching and why? The simple answer is life. The soil is alive and this life needs nourishment. There are more living things in a handful of rich soil than there are people on the earth and the only way they can survive is if they are regularly given something to eat. The organisms that inhabit soil are many – and while the bulk of soil is tiny fragments of weathered rocks and minerals – there are a wealth of other components: plant matter, dead animals, and then the group of organisms that continues the decay of those things (bacteria, fungi, viruses, and so on) until we come to see that this huge mix is one continuous cycle of life and death, the by-product of which is food for plants in the form of humus.

The soil can render death and decay into life-supporting food for plants. Consider then for a moment how much soils need to eat if they are responsible for plant growth as well as their own nutrition. Lots is the answer, and continuously through all four seasons of the year, even if plants are dormant in the cold months of the year. This is why the use of mulches is so important – they protect and feed the soil to sustain its life.

As an example of what lengths we might need to go to in order to get hold of a supply of mulch, I cite Narayan Reddy, a well-known organic farmer from just outside Bangalore in the state of Karnataka in south India. One of the ways in which he urged farmers and growers to come by organic matter was to collect leaves and animal droppings off the roads to make compost. In south India, the prevalence of public animal waste is obvious, India's being a largely agrarian economy. Although we no longer have such an economy in the UK, this example should encourage us to be resourceful and ask farmers and stable owners to part with spoiled straw or manure. Some will try and sell it to you, but usually you can get it for free because they are keen to get rid of an ever-increasing heap. We should approach our local councils too. Leaves in an advanced state of decay are constantly being swept up from roadsides by councils in autumn on their way to landfill. We could do worse than collect them ourselves and make leafmould.

MULCH continued...

of straw which, like wood chips, is high in carbon, but the nitrogen-rich urine contained in manure will help balance the carbon content and prevent this mulch from denuding the soil. Straw alone is good as a mulch, but avoid hay because it tends to be full of grass and other seeds that may germinate in the soil. Homemade compost is best (even though the top layer might dry out in hot weather) because it adds nutrition, retains moisture and will be taken in by the soil. While bags of organic compost sold at garden centres are expensive, they may be your only option. Lastly, seaweed makes excellent organic matter, if you have access to a supply.

COMPOST

What I am going to concentrate on here is the manufacture of compost in our gardens and on our own smallholdings. Specialist compost mixes used for propagation and sowing seeds will be discussed later because it is important when growing plants from seed that the compost is sterilised against pathogens and is not too high in nutrients.

What we are talking about in terms of compost is the material that we create from household and garden waste for returning to the soil. This is the most important type of recycling because it provides food for our plants, which would otherwise have to be brought onto the site and costs money, energy and time. The three main elements that plants need for healthy growth are nitrogen, phosphate and potassium. Compost made from a good mix of garden plants and kitchen waste will provide these elements in balanced quantities.

Composting is a very straightforward process and starts with a positive – 'compost wants to happen' as Elliot Coleman says in his book the *The Four Season Harvest*. Organic matter will decompose and turn into soil under its own steam. If we are to accelerate the process (because we need the precious material for our gardens or containers as fast as we can produce it), we have to be sure that we are adding the right materials to the compost mix at the right time and in the right condition. By making compost, we are mimicking the natural process – organic matter gathers on the forest floor and breaks down gradually to form soil. But that process is slowed down when a large tree branch falls to the ground and starts rotting. In the garden, compost can be matured from scratch in three months, winter or summer. The key to the speed of the process is the size of the material that you add to the heap.

Most kitchen waste may be added without fear but, if you live in an urban dwelling, it is probably best to avoid adding leftover bits of meat to the heap, which may attract the attention of rats or foxes. I have never followed a 'no meat' rule. If you have an active compost heap, the material that you add will break down very quickly. Vegetable peelings, tea bags, leftovers of all kinds – pile it on because it will all rot down in the end. It came from the soil and was nourished by it in the first place and so will eventually return to it.

Garden waste will make up the bulk of the compost heap. General maintenance of your productive garden will produce hedge clippings, lawn mowings, old compost from pots, dead and dying plants, weeds, soil and organic matter of all shapes and kinds. It can all go on the compost heap, but there are three points to remember. The first is that the material is not too woody because it will be harder to break down, the second is that it is chopped up small so that it is all touching (to allow the organisms that are responsible for the breakdown process to function at their best), and thirdly that it is spread evenly over the heap rather than dumped in one lump, which will hinder the breakdown process.

There is no need for turning, watering or covering or adding any compost heap 'activators'. The heap will form its own skin and protect itself from the influence of the weather and break down naturally. The contents of the heap will start the process of decomposition without needing any encouragement from external factors.

Instructions for building compost heaps can be confusing because they usually assume that you have an enormous quantity of material from the outset. They suggest that only a large heap will build up enough heat to kill weed seeds and pathogens in the compost, rendering it safe to use on your growing beds. Don't worry about this. It is much more likely that you will have small quantities of material to add to the heap – a daily bucket of fruit and vegetable peelings from the house and a barrow or two of weeds and plant material from the garden per week. As such, the heap builds slowly and, while it may not heat up to the maximum, this will not stop the breakdown of its contents. Also, any fungi or bacteria transferred from the newly made compost to the soil will do little or no damage. The lack of heat in a smaller heap may not kill weed seeds, but every soil is full of annual weed seeds and they are going to grow anyway. For weed control, you need to keep your hoe sharpened and in regular use, or your soil covered with mulch to suppress the weeds.

The top of the heap – being the coldest – will take the longest to break down, but after three months, when you have removed the two-thirds of the heap that has turned into compost, put the uncomposted top section to the bottom of the heap and start all over again.

For most people, a total of two compost heaps is adequate to serve the needs of a family garden. Compost bays are easy to construct and should be about 2m (6ft) in both width and depth with a height of about 1m (3ft) for the easy transferral of green waste from wheel barrow to heap. Wooden pallets (which are widely available secondhand) make excellent materials for compost bays because the wood 'breathes' and the pallets raise the base of the

COMPOSTING WEEDS

When you are starting a gardening project and you have lots of perennial weeds to dig out and remove from the site, you must be careful that, if you compost these weeds, you provide means by which they can rot down properly. If they don't and you add only 'half-cooked' compost back into the garden, the perennial weed problem will arise again. The way around this is to build an entire heap from scratch, and this should be hot enough to kill off the weed seed. For maximum efficiency of the breakdown process, do not add huge lumps of perennial weeds on to the heap all in one go. Instead, layer the heap with some of the weeds and then a high nitrogen layer of manure or grass cuttings. Add wood ash to boost potassium levels in the compost, as potassium is water-soluble and tends to leach out.

compost off the ground. Chicken wire is also a good material, since it allows for proper air circulation, but you may have to buy it new.

If the construction of a compost bay is beyond you, there are various compost bins available to buy, some of which are donated free by local authorities in an effort to encourage composting of household waste. These compost bins tend to be smaller than a homemade bay and their conical or tub shape does not seem to encourage the breakdown process quite so efficiently.

For the small space, a worm composter is the best choice. Vermiculture mimics nature by assembling a larger concentration of worms than would normally be found in a given amount of soil and then setting them the task of eating the organic matter offered. It is an excellent way of composting household and other green waste that is produced from a balcony or a terrace garden.

The recycling of nutrients around the garden via the compost heap is gratifying, but what is particularly satisfying is knowing that the finished product will have an even content of nutrients (assuming that you have added a wide range of matter to the heap) to suit almost all the plants that you grow. The site for the compost heap is not vital, but a shaded area keeps direct sunlight away and prevents the heap from drying out.

The raw material in its uncomposted state will look quite bulky. Sadly, as the process continues, the heap will shrink and the end matter is considerably less than the original. How, then, can we best use the compost that has been made? The application of compost depends on your gardening priorities but, generally, compost as a complete plant food is excellent for laying on the surface of beds and enriching soil, for mulching anywhere, or for filling containers in which to grow plants of all types and varieties.

Right: Compost: the lifeblood of the soil and the redemption of the garden.

FERTILITY

With soil in good heart, compost for mulching and feeding, and a rotational plan, you are well on the way to building a fertile soil. Keeping the soil in a fertile state is a continuous process and is the prime concern for the 'sufficient' gardener. Soil and plants are both hungry and in constant need of care and attention. However, as I have mentioned before, if you ask of your plants and soil only what they can give – not what they must give – you will be successful in maintaining a healthy balance.

If you build your garden on a foundation of fertile soil and nourish it, you will produce healthy plants and have fewer problems with pests and diseases from the outset. A well-stocked garden environment can be a model of bio-diversity and can contribute to the balance of nature. It is not a question of having specific predators to deal with specific pests – too much is made of this. It is all about balance, and this is what bio-diversity is: the equilibrium of life.

There is an argument that puts forward the idea that a pest or a disease will continue to target a weak plant because it is not getting what it needs. If it feeds off a strong plant, it will do so and move on, satisfied. For example, healthy roses often appear to be hosting vast quantities of sap-sucking aphids, which then seem to disappear almost as soon as they arrive. This does not mean that the slugs and snails will disappear from your garden overnight, but it does mean that your healthy plants are less likely to suffer from infestation from the lower orders of pests that are present in an unbalanced environment.

With a fertile soil as the main aim of the gardener, we must explore the ways that exist to fertilise the soil directly. The use of compost and mulches is very important, but I also suggest a fertilising regime for each crop, over and above strong soil fertility. The rotational system of annual crops where fertiliser is added manually, and the use of leguminous crops to fix atmospheric nitrogen, has great value and we should also look at green manures to enrich the soil.

Green manures can be any number of annual or perennial crops that are sown directly into the soil before being dug in at a later date to provide bulky organic matter for the soil life to break down and turn into available nutrients. Some, such as clover, have the added benefit of being leguminous, while others, such as *Phacelia tanacetifolia,* are strong insect attractants. If you are growing large amounts of annual crops, a green manure in the rotation over winter is a valuable use of space and will prevent erosion and hold soil structure.

Left: Treat it with care, compost has a life of its own.

COMMERCIAL VIABILITY

The purpose of this book is not to turn the 'sufficient' grower into a commercial producer of organic fruit and vegetables but rather to show how you can contribute to your family and community life by producing zingy fresh salads and high-quality organic raspberries full of flavour.

The pleasure and satisfaction of feeding your family and friends may tempt you down the road to becoming a full-time organic market gardener. Such a life, whilst enormously rewarding – and I know because I have been one – is extremely hard work. My advice is to view your kitchen garden as a wonderful and productive way to spend your free time.

When you are overwhelmed by a bumper crop, see this as an opportunity to exchange and barter with your neighbours or as an opportunity to give away fresh produce to those who do not have. And, of course, throw a large party and boast proudly of growing everything on offer.

There is a dearth of smallscale market gardeners throughout the Western world and whatever way we can spread the idea of 'sufficient' will be positive. A fruit and vegetable stall with an 'honesty box', offering fresh produce to passers-by, is an excellent way of involving people in the 'sufficient' activities that take up your time and thinking.

Making your activities known to your community might help you find a source of compost and, in return, a home for excess vegetables. Let people know that you are gardening not only for yourself but for the betterment of community. This is what 'sufficient' gardening is all about, gardening for you, for wildlife, for the environment and for others. It has to be inclusive, or it does not work. Nature works on the basis of generosity and abundance, providing enough for her plants, trees and animals, and so must we.

Right: A roadside stall: any surplus produce for sale will always find a home.

VEGETABLES

GROWING FOOD

So much of growing food comes down to personal choice and the ability to put that choice into practice, taking into account the amount of space you have, and the time and energy you can invest. That it is the best of fun is not in doubt, but there are no end of crops that are frustrating, difficult, fussy and generally a nuisance to try and grow. I am going to point these out along the way to try and prevent the heartache of a huge effort for what might turn out to be a poor return. There are some crops that are good for storing and some that will save you a great deal of money. There are some that are seasonal luxuries, and others that taste far better when they are homegrown and eaten fresh. There are also plants and crops that will be of benefit to your garden and your soil. The choice is vast but, once again, a balance is what is required to make sure that we have 'sufficient'.

For some crops, it is much more rewarding to have someone else grow them for you. For example, it is delicious to grow a handful of early or late carrots but, if you want enormous quantities of beefy main crop carrots for juicing, it is almost certainly more time- and cost-effective to buy them in, while you use the space to grow something else. A great many of the brassicas come into this category, and I will not hesitate to flag these up too. Yes, failures can be composted or fed to animals, but that is not the idea. What I want from my food-growing space is a delicious and nourishing vegetable or fruit that has not taken me sleepless nights and a lot of unnecessary hard work to produce. One that has been enjoyable and interesting to grow and that fills the senses with the pleasure of growing food.

Take a plum tree, for example. What does it offer? Scent, beautiful blossom, insect and bird activity, soil stability, leaves for leaf mould, shade and, of course, fruit. Fresh locally grown organic plums are hard to find anywhere and, as such, can be considered a luxury. Main crop potatoes, on the other hand, are not a luxury; rather, they are a staple. Diet and taste depending, many would cite them as a crop that they would need to grow in some quantity if they were to be self-sufficient. Ten 30ft (9m) long rows, 3ft (1m) apart would feed a family of four through the year. That equates to 900 sq ft (83sq m) of space required for main crop potatoes. It's the size of a small back garden and therefore completely out of the question for many of us. Organically grown main crop potatoes are relatively inexpensive and available. They take a lot of growing given the prevalence of disease such as blight, and require a heavy dressing of organic matter. Their yield, beyond the tubers themselves and some which can be saved for seed, is an amount of plant material for composting and the resulting well-fertilised, weed-free ground. Nothing else. You could probably barter your plums for potatoes, but I know which crop I would rather be growing. Let someone else do the work.

This is the point, that an attitude of 'sufficient' leads to a style of growing and a type of gardening that becomes enjoyable rather than a chore. The state of self-sufficiency is a tall order; the state of 'sufficient' is more about understanding and appreciating what can come from the land rather than what must. Any connection with the land is rewarding, but particularly when you are co-operating with it to produce wholesome crops for yourself and your family.

PROPAGATION

The range of crops we can grow, from the tiniest salad plant to the tallest tree, is enormous and the different methods of cultivation far-reaching, but this hybrid model that is called 'sufficient' combines both traditional and state-of-the art methods of food growing. At its heart, however, is a discipline that is common to all forms of plant production – the raising of a plant to have a strong root system that will grow well once planted. It is important that plants start life in the best possible condition to give them the best chance of being productive and staying healthy.

Once you start gardening, you'll find that plants arrive in all conditions and in all different stages of development. They may arrive as a packet of seeds, a leftover tray of seedlings from a neighbour, a division of a herbaceous plant from someone's border, or as a plant that we buy from a nursery. We have to know how to treat these plants on arrival and understand their nutritional needs before they can be transplanted out into the garden.

It is essential to have a significant quantity of compost on your site at all times in order to pot up new plants. Ideally, this material will be homemade but, if not, an eco-friendly compost, such as coir (the name given to coconut fibre that has been processed into compost), will do.

SEED PROPAGATION

The propagation technique used depends on the plant in question. There is no single way for all. When we raise plants from seed – this is the method by which we will raise almost all of our annual plants – we must use sterile compost to avoid disease. Tiny seedlings in pots are at risk from pathogens in this confined space and, just like newborn babies, they need protection. The only exception is when we sow seeds for propagation in open ground as they are less prone to attack.

After germination, the seedlings are left to grow on until they are large enough to be moved into a growing space of their own. This moving process is known as 'pricking out', and here we have to be very delicate with the seedling, as it is at a very vulnerable stage of its life, handling the seedling only by the cotyledons (juvenile leaves).

Now comes the point where you can use your own homemade compost as the material into which the seedling is 'pricked out', as long as the compost has been sterilised to remove pathogens. It is simple to do yourself, requiring only a little time, and will save a great deal on buying expensive and energy-inefficient bagged compost. All you need is a small camping stove and a large saucepan, in which you quite literally 'cook' the compost until it is hot to the touch and then leave it to cool before use. This process will not harm the compost, it will only kill the pathogens.

If the plant is growing fast and needs potting on again into a larger container before being planted out, it can be potted into homemade compost and will benefit from this new source of nutrients. If these rules are adhered to you will have healthy plants ready to hit the ground running.

During the propagation process, the temperature and light requirements of plants raised from seed are particular to each plant (see each separate plant heading), but the watchwords are 'airy, light and cool', whether you are using your kitchen or a greenhouse. Warmth is good too, but go steady; extremes are to be avoided.

Finally, the action that damages plants more than any other during the propagation process is overwatering. There are no strict rates of application, only that the roots need to be able to take up moisture from a compost which is moist but not too wet. It is something you learn as you go along, but waterlogging is undesirable and will kill the plant, so again, go easy; you can kill plants with what you think is kindness.

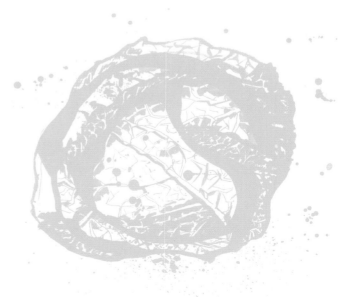

SOLONACEAE

POTATOES, TOMATOES, PEPPERS, CHILLIES, AUBERGINES

POTATOES *(Solanum tuberosum)*

After rice, maize and wheat, the potato is the most widely grown field crop. On that basis, it is a staple, particularly outside of Asia, where rice holds sway. As a tuber, and therefore grown underground, the potato has considerable benefits over grains. It is far less susceptible to aerial threat from birds, and less damage is caused by heavy wind or rain. It is, however, frost-tender and prone to a crop-killing fungal disease called potato blight (*Phytophtora infestans*), which is endemic in parts of the world where warm, moisture-laden air flows dominate, in particular the UK and Ireland.

HISTORY

The potato started life in the Andean regions of Peru before journeying to Europe in the sixteenth century with the great explorers. Finding similar growing conditions in Europe, it has been a staple of diets all over the world because it is a 'complete' food – one which we could survive off if we ate nothing else. The potato has value as a source of carbohydrate and, to a much lesser extent, protein, but it must be eaten cooked to break down the starch content of the potato and make it digestible.

HOW TO GROW

The humble spud is a very amenable subject in the garden and, apart from frost and blight, it presents few problems for the grower. It has a few likes and dislikes. It likes the sunshine, not the shade and it needs space to grow. It is both hungry and thirsty. Without good quantities of food and water, it will not thrive. In the era of climate change and potential drought, the potato may not have such a bright future commercially, but for the 'sufficient' grower a small quantity is easily attained, and as a crop that stores well it has added value.

The traditional way to grow potatoes is in long straight rows, which are ridged up to control weeds and give space for the tubers to swell into. The 'ridging up' is labour-intensive, and while it will control weeds, it is just as easy to bury a seed potato tuber under a mulch and arrive at a good harvest of potatoes without interfering with the soil at all. Potatoes are good subjects for containers too, big ones such as dustbins, or stacks of rubber tyres if you insist; they like the space in which they can send out their roots (extended stems that grow underground), on which the tubers swell. The smell of newly lifted potatoes when they appear from the soil is incomparable and the squeak they make as they rub together when handled is almost human. We should all make room for some new potatoes (first and second earlies), but I strongly recommend leaving the production of bulky maincrops (which occupy a lot of space in the ground for quite some time) to others.

SEED

In early spring, seed potatoes are dispatched to the customer from seed suppliers and become available to buy at garden centres and country stores. It is a good idea to get them 'started' before you plant them to speed up their young growth so that they crop earlier. This is a process called 'chitting', whereby you encourage the tubers to sprout small leaves. Set the seed potatoes upright in a light, airy and frost-free place indoors with the 'eyes' – small slits from where the sprouts appear – pointing up like a smiley mouth and the 'tail' of the potato – a wisp of dried root – facing down. When the small leaves have put on about 1cm (⅓in) of growth, they will be ready for planting. This will give the seed potatoes about five to six weeks underground before the dark green leaves of the potato plants start emerging through the soil. The timing should coincide with when the last frost has passed.

SOIL AND SITE

Access to full sun is essential to potatoes; even part shade will result in leggy foliage and poor tubers. What they also need is food, in quantity. They are not fussy about soil types although chalky soils will result in the appearance of scab – a fungal disease that shows unpleasant scabs on the surface of the tuber but does not affect the flesh. Potatoes like a rich soil holding plenty of organic

matter. They also favour potassium, and as this is an element that is highly water-soluble (i.e. it washes out of the soil quickly), it is good to spread as much wood ash, which offers a good source of potassium, on the potato patch as possible, or to mulch it with the potash-rich herb comfrey.

PLANTING

Allow at least 45cm (17in) between plants, maybe 60cm (23½in) for main crop potatoes, and at least the same between rows so that you can ridge them up if you wish. My preferred method would be to plant through a mulch about 10cm (4in) deep with the 'chitted' sprouts facing up. There is no need to water; simply cover up the tuber in the hole with the soil and gently firm it down with your hands. For early varieties, 90 days is the minimum before you can start harvesting; for main crops, 110 days.

VARIETIES

Thousands of different varieties of potato exist all over the world and they fall roughly into three groups as far as maturing times are concerned: first early, second early and main crop. Aside from that, each variety of potato is different and may differ in flavour, texture, size, shape and colour and, perhaps most importantly of all, they differ in dry matter or moisture content, which determines how they are best cooked.

For the 'sufficient' grower, I would recommend growing a number of first early potato varieties, which mature quickly, and perhaps an early salad potato variety for their waxy texture and strong flavour. Growing earlies means that the blight fungus, which appears later in the growing season, can be avoided. For me, early or 'new' potatoes are a magical crop to be savoured for a few weeks of the year, as they don't store well. There is nothing more delicious than the first new potatoes boiled and served with butter and fresh parsley – they are one of life's great luxuries. Thereafter, if you want potatoes for roasting, boiling, mashing, baking or for chips, I think you are better off buying them in. Even though there are many maincrop potato varieties with blight resistance bred into them, the risk and the effort is too great. Use the space in your plot more wisely and leave the growing of maincrop spuds to someone else.

When the potato crop is growing, keep an eye out for slug and snail damage on the foliage and hope for good rainfall because it is rain that swells the tubers as much as sunlight and food. If you have only a few potato plants and there is a drought, it is probably not worth watering them – the water is better used elsewhere on smaller more delicate plants.

At harvest, make sure the potatoes are dry before storing them as quickly as possible in paper or hessian sacks in a cool but frost-free environment. Do not leave the newly harvested potatoes on the soil surface, as exposure to the light might turn them green and toxic. The foliage, or 'haulm', can be removed immediately to the compost heap as long as it is free of diseases, such as blight. Once potatoes are in store, rodents are the problem. Make sure that the potatoes are out of reach of rats and mice by suspending the sacks off the ground, or by storing the sacks in dustbins.

The Potato Famine

The Irish Potato Famine of 1847–51 came about for a number of reasons. The first was lack of diversity in the crop variety. The second was poor husbandry on holdings that were too small to grow anything other than a monocrop of potatoes to sustain the family, and the third was the general presence of the blight fungus in ideal climatic conditions in Ireland. The famine killed more than one million people and was one of the greatest tragedies of the nineteenth century, supposedly an era of expanding affluence, wealth and technology. As the fungus rendered first the foliage and then the tubers black and stinking, so the population starved since little else was grown and help from the apathetic English was slow to arrive.

Right: Tomatoes are a simple and reliable crop for the 'sufficient' grower with a multitude of uses.

TOMATOES

The second most widely grown member of the *Solonaceae* family after the potato, but perhaps the most important for the 'sufficient' grower, is the tomato. This is a plant that produces fruit of all shapes and sizes and has versatility as its strongest suit. There are varieties for all climates and situations, among which are cookers, eaters, beefsteaks and some as small as raisins. Colour range and flavour are diverse but, like the potato, tomato plants do not care for frost, they appreciate sun and require potassium to produce high-quality fruit. Tomato plants need physical support, as they grow either bushy and heavy, or tall and lanky, with no means of supporting themselves. They are easy to grow, but the plants need attention at certain key stages of growth to give of their best. Any good garden soil is fine for growing tomatoes, but if you wish to plant them in containers, you will need a plentiful supply of homemade compost and a supplement of liquid tomato feed to produce a healthy crop.

Tomatoes are well-suited to pot culture and because of the fungus blight – which appears late in the growing season when outdoor-grown tomatoes are still ripening – they are best grown in the house on a sunny window ledge or in a polythene tunnel or glasshouse. In hot climates – where incidences of blight are fewer – they will flourish outside in open ground. In truth, the best time to eat tomatoes is in the height of summer, when the flavour is at its most intense. The bouquet when they are freshly picked is quite out of this world. Storing fresh tomatoes is difficult and so, if you have a large harvest, they are best cooked and bottled in airtight jars or made into delicious fresh tomato sauces for freezing. They also make wonderful relishes and chutneys. If you have a polytunnel or greenhouse, the tomato growing season can be stretched out into the 'hungry gap' after Christmas, but by then the flavour will have fallen away.

SOWING AND GROWING

As they have a long growing season, tomatoes need to be started early and with some heat. Fill a 7cm (2¾in) or 9cm (3½in) pot with a sterile organic seed-sowing compost to just below the surface. Place about 10–12 of the little furry seeds on the surface of the compost, without allowing them to touch one another. Sieve a little compost over the top so the seeds are covered, and place the pot in a warm, light and airy place that has a reasonably constant room temperature. When the seedlings are 4cm (1½in) high, they are ready to prick out. Transfer each one to its own 7cm (2¾in) or 9cm (3½in) pot filled with

sterile compost and grow on for a further two weeks, still in the heat. Two weeks on, and they will be ready to pot up into a 1 litre pot of homemade or loam-based compost, from where they can be moved into their final home outside as soon as the chance of frost has passed, or undercover anytime.

The two types of tomato are the cordon and the bush. The cordon grows as a single-stemmed plant, with the flowers and fruit borne on trusses that protrude from the main stem. These plants must have every side shoot that appears at the axil between the main stem and leaf removed, as the plant grows and they will need tying into a cane to support them as they grow taller. Bush plants are allowed to grow freely, and fruit will be borne on the multitude of stems that appear from the side shoots and the growing tip at the top of the main stem. It sounds complex, but with cordon tomatoes the side shoots that appear in the angle by the leaf and the stem have to go – they are a whole new tomato plant in the making and will create havoc if forgotten. It is critical not to mistake the side shoots for the flower trusses, which are produced on the main stem roughly halfway between each set of two leaves.

If you can't raise your own plants from seed, ask around to see if friends have surplus young tomato plants or buy them from shops or plants stalls at fairs. Once your plants have arrived and have 6–8 true leaves, they are ready for potting into their final place. Choose a suitable container unless you are intending to grow the tomato plants directly in open border soil in a glasshouse or a polytunnel. The minimum size of container should be about 20cm (8in) across and 20cm (8in) deep. Homemade compost is a good growing medium because it has a balanced content of nutrients and it drains freely. (This is because compost breaks down over time and, no matter when you use it, there is always a quantity of material that has not broken down fully. Also, there is always a high content of stones and sand that help it drain). You will probably get annual weed growth on the surface of the compost in the pot. Pick off these weeds and put them on the compost heap. Alternatively, leave the picked weeds on the surface of the pot to shrivel up and rot down and act as a mulch.

Before planting, make sure that there are drainage holes in the bottom of the container. To aid drainage, these holes need to be loosely covered with bits of old broken clay pot or stones. Next, fill the pot three-quarters full with compost and insert a stake to support the plant – driving a stake through the roots once the tomato has been planted is not good husbandry. The stake need not be very tall, perhaps 45cm (18in), because ultimately, the best way to grow tomatoes is up a string once they have grown taller than the cane. They are not climbers, but they grow very tall and the main stem likes to be wrapped or tied around string, as it is more flexible than cane. The vertical strings need to be tied to a horizontal support in the roof of the growing room to hold them upright. This rigged system needs some planning and design in advance.

Tomato flowers need pollinating to produce fruit and this work is done by insects; bees, in particular, do a lot of the work. If you are growing tomato plants indoors or under glass, open the windows or vents during the day to allow insects in. Close them at night. Tomatoes prefer a dry growing atmosphere but will still need watering morning and evening.

GOING AGAINST THE GRAIN

Conventional wisdom has it that only 6–8 flower trusses should grow on any one tomato plant before the growing point is 'nipped out' to prevent further growth. (This method aims to ensure that all the tomatoes produced will have time to ripen.) However, as long as the plant remains healthy there is no reason why you should not allow it to keep on flowering up until the end of the growing season. As long as you keep on removing the side shoots (see above), you can allow the plant to continue growing. Tomatoes that do not ripen on the plant can always be made into green tomato chutney.

FEEDING

Tomato plants grown in containers need feeding because the compost used in planting will last as a source of nutrients only for two to three weeks. After this, a tomato plant will need feeding once every two weeks. A mulch placed around the base of the stem of the plant and covering the surface area of the pot is also beneficial. The best way to feed tomato plants is to use a liquid feed as a root drench. Once every two weeks, replace your evening watering with a liquid feed watering instead. Comfrey leaves are rich in potassium and nitrogen and make a good homemade liquid feed for tomatoes, potatoes and the other members of the *Solonaceae* family and in fact most vegetable plants.

COMFREY LIQUID FEED

The variety of comfrey (*Symphytum* x *uplandicum*) known as 'Bocking 14' is an herbaceous perennial with a higher content of both potassium and nitrogen than both homemade compost and farmyard manure. It has been favoured by the organic growing community in Europe for many years since Lawrence D. Hills, the founder of the Henry Doubleday Research Association, crossed species of comfrey to produce a very vigorous and nutrient-rich variety, which he called 'Bocking 14' after the trial ground where he did his experiments in eastern England. It is a non-invasive comfrey, unlike other members of the Symphytum genus, but the plant does grow very tall, up to head height if well fed (even plants that make feed need feeding!) and, in the process, produces plenty of bulk for either composting, mulching or for making liquid feed.

To make a comfrey liquid feed, pack as many leaves and stalks as you can into a water butt with a tap fitted in the bottom. Press down the leaves with a weight so that they compact together. Raise the butt off the ground on blocks so that your watering can is able to fit under the tap. After two weeks, the compacted leaves will produce a liquid and the feed will be ready to use. It can be released through the tap and diluted with water at about ten parts to one. When no more liquid is produced by the comfrey leaves, they can be taken out of the butt and put on the compost heap – where the leaves will stink mercilessly until they are covered up by other compost material and rot down.

TROUBLES

Tomatoes are one of the least neurotic crops and are best grown in what is known as 'protected cropping' – i.e. under cover. This, together with their wonderful flavour and high nutritional content, makes them a favourite with the home grower. However, growing a quantity of any plant indoors in the warm is bound to lead to the odd problem because we are going against nature. Here, we have to be super vigilant and check the plants regularly to prevent infestations of pests, such as whitefly and aphids. These pests are problematic for tomatoes and are best dealt with by spray-misting the plants with a weak dilution of an ecological washing-up detergent. (The oil in the detergent will immobilise and kill them.) Check the growing tips of the plants and the undersides of the young leaves, as the nitrogen-rich sappy growth of the plants is attractive to the pests. Alternatively, introduce a biological control, such as the parasitic wasp *Encarsia formosa,* which deals effectively with whitefly. You are unlikely to have disease problems with tomatoes, but if the bottom of the fruit is marked by a brown spot (about the size of a penny piece and known as blossom end rot), your plants are deficient in potassium and you need to review your feeding programme.

SAVING SEED

If you have success with a particular variety of tomato, try to save some seed so you can grow them again the following year. Saving seed is something all 'sufficient' growers should do, not only to preserve a particular variety but also to keep costs down and reduce energy expenditure. Seed can be saved from numerous different vegetables, herbs and fruits, and by doing this we not only increase our stock of plants but we can replace worn-out plants and build a seed bank. It takes little space and only a modicum of knowledge to start the process, and once you get going you will never look back. Once you get the techniques right, it becomes absorbing.

As to the choices – seed potatoes are an easy one, as you have simply to choose a tuber the size of a hen's egg or a pingpong ball and keep it somewhere cool and dry over winter. Brassicas are easy too. Salad rocket, for example, produces a seed case bursting with tiny brown seeds after the white flower has disappeared. All you have to do is to time the collection right to prevent the seed pods from exploding and covering a wide area. With the price of seed spiralling and the concerns about contamination via genetically modified organisms and chemicals, you know what you are getting if you save your own seed (or maybe you don't, because if two varieties accidentally cross you might breed a new variety!). Encourage your friends to save seed and set up a seed-swapping bank. Grow to swap.

SWEET PEPPERS AND CHILLIES

I regard both sweet peppers and hot chillies as luxury crops, largely because they are not heavy bearers of fruit like tomatoes. There is a wide choice of seed available from specialist growers, which offer a more interesting range of flavours, shapes and colours than the sweet peppers and chillies that you can buy in the shops. Both peppers and chillies are easy to grow – the hardest part is germination because it takes time and the tendency is to panic when the seedlings don't appear. It is easy to do something foolish like overwater them or move the seed tray to a warmer position. Like tomatoes, these fruits are not troubled much by pest or disease. They grow well in even heat and plenty of light, and as with all members of the *Solonaceae* family, their nutrient of choice is potash.

It is easy to become passionate about peppers and chillies. Once you have grown your own successfully, it feels like you have made a breakthrough on the 'tough things to germinate' list. Although I have said that they are easy to grow, both sweet pepers and chillies have a way of confounding the grower by not doing the obvious thing at the right time. You need to keep cool with these plants, as they often do what you don't expect. Many have terminal buds, by which I mean that the stem splits into two branches once it reaches 45cm (18in) tall, and it appears to stop growing. Then when the small white flowers appear they take several weeks to turn into fruit. Once the fruit starts growing, you will not believe that it is capable of reaching maturity. However, by the end of the summer and into autumn, you will have magnificently shaped, colourful fruit that bear little or no resemblance to anything you have ever bought from a shop.

SOWING AND GROWING

Like tomatoes, peppers and chillies need to be sown and grown in the warm. I see no point in being fussy about exact temperatures; you are hardly going to alter the temperature of your house for the sake of a few chilli plants. A warm kitchen windowsill is perfectly adequate for sowing. In 10–12 days you should see germination, and in another week, decent-sized cotyledons (primary leaves) should have appeared. These seedlings are then ready to be pricked out into 7cm (2¾in) or 9cm (3½in) pots. After a further two weeks, they can be potted into their final container or into a growing bed in the greenhouse. The compost used and the growing conditions are similar to tomatoes – homemade compost is fine because it is nutrient-rich and has adequate drainage.

Feed the plants well fortnightly with a high potash feed and stake them with canes for support because the fruits are heavy. When you have harvested the last sweet pepper, cut off two-thirds of the plant. Keep over the winter in a warm room for a period of self-imposed dormancy. Simply forget about the plants until early spring. The plant will grow and fruit again next year because in their native habitats, such as South America and Asia, sweet peppers and chillies are perennial shrubs.

To dry chillies, simply lay out the fresh chillies on newspaper and allow them to shrivel up until all the moisture is removed. Store them in a dry environment; otherwise, I suspect that they may rot.

AUBERGINES

The aubergine, also known as the eggplant, presents more of a challenge to the grower. Chillies and peppers are good to try because you can always use a chilli, fresh or dried, and peppers are a must-have luxury. Aubergines come into a slightly different category in that they are a rarely used vegetable, but very tasty if you can get them to grow. In the glasshouse, where space is always limited, I would concentrate my efforts on tomatoes, one or two plants of an all-female hybrid cucumber (to ensure a good quantity of bitter-free cucumbers), a few sweet peppers and chillies but leave out the aubergines. (Some might argue that melons should be grown inside, but I think these are better off being allowed to roam wild in a hot spot in the garden, where they will cross-pollinate more successfully and hopefully produce more fruit. All shop-bought aubergines look fantastic, but because they are imported from the Mediterranean and south Asia, they are expensive and rarely taste good. The truth of the matter is that aubergines need hot weather and the place for them is not the cool temperate zone of western Europe, even if you have a hot glasshouse.

If you must grow them, aubergine plants have the same requirements through the growing cycle as tomatoes and chillies. They like the same compost, nutrition and watering but favour constant temperatures both day and night, and have a longer growing season. If your glasshouse cools down at night, growth will slow considerably and this will have a bad effect on productivity.

The key with all of the crops in the *Solonaceae* family is to remember that they need potash and like nitrogen. Compost will give you both, but more in the way of feed is always needed because this family is a greedy one. When you are setting up your productive garden, make sure to establish a bed of comfrey plants (see pages 95–96). Once you have done so, you will have it for life. It can be in a shady corner so as not to occupy a valuable sunny spot, but remember to feed it by mulching it down with manure or compost in the winter. And remember also that the glasshouse-grown members of the *Solonaceae* family like heat, so shut the greenhouse doors at night, whatever the weather.

BRASSICAS

The brassica tribe is enormous. Its spectrum covers the basest of animal fodders to some of the finest tasting vegetables. What many people know as 'English' vegetables – cabbage, cauliflower and Brussels sprouts – are out of fashion as popular tastes have moved towards such delicacies as purple sprouting broccoli, sea kale, salad rocket and many exceptional salad greens. Ironically, many of these more popular greens are also members of the Brassica family. Other members include turnips and swedes (known as rutabagas in the US), kales, Chinese cabbages, calabrese, mustards, rape, radishes and even a stem brassica by the name Kohl rabi, which is especially popular in Germany. Last but not least come the seeds of many brassicas, which can be sprouted and eaten as sprouting seeds or baby salad leaves. These and the numerous types of exotic brassica leaf used raw in salad (and sometimes cooked) are excellent for growing in containers. Cabbages and Kohl rabi can also be cultivated in containers. However, the bigger brassicas (Brussels sprouts, cauliflowers, kales, calabrese and the sprouting broccolis) put down a long tap root and grow best in open ground.

Brassicas are willing growers through most stages of their growth cycle, but to produce the best results, we have to understand them. They may appear ordinary and seem somewhat dreary winter fuel, but they are sensitive and have foibles, which makes them fun and a challenge to grow. What is more, brassicas are one of the few sources of fresh greens through the winter. Seasonal food is a very important element in the concept of 'sufficient' and if you intend to follow this path, brassicas can make up a significant part of our winter fibre.

The real secret to enjoying brassicas lies in the cooking; overcooked greens experienced as a child is the source of many peoples dislike of brassicas. When cooked well, the flavour of homegrown and freshly cut greens far outweighs the shop-bought equivalent, and they are far richer in vitamins and nutrients. In truth, there are some brassicas I would grow and some I would leave out simply because they are hard to grow well. We should remember that growing takes time and effort, and if the results are poor we may lose heart.

In terms of cultivation, brassicas like their ground to be firm. They do not

'Cabbages and kales are the most straightforward of the winter brassicas to grow, having little in the way of specialist requirements, needing only a reasonably fertile soil.'

like freshly manured ground but ground that has been manured the previous spring for the crop (usually potatoes) that preceeded it in the rotation. Temperature does not bother brassicas; they can withstand great heat and extreme cold, but they must have a well-balanced soil in good heart. Birds, slugs and snails are a problem, as are the larvae of certain butterflies and moths. Brassicas can also be troubled by smaller pests in the way of aphids, whitefly and cabbage root fly and finally, if they are grown in the same place year-after-year, they can fall foul of a brutal fungal disease called club root, which persists in the soil for many years. The only way to be rid of it is to rest the ground from brassicas for seven years.

Despite their vulnerability to pests and the fact that some brassicas (brussels sprouts, for example) take up a lot of space and can be inconsistent in their growth, brassicas make a good homegrower's crop. They are labour-intensive but they come into their own in the 'hungry gap' between late winter and early spring, when the garden is at its least productive and there is little else to eat. Some, such as calabrese (green-headed broccoli), make excellent summer crops, and because it is not hardy, this is the only time you can grow it. If you have ever had trouble getting your children to eat their greens, try involving them in growing their own calabrese. When they have been part of the growing and cooking, the eating will follow. My belief is that the big brassicas are best grown as winter crops, with the exception of calabrese and the odd summer cabbage, which can be shredded raw for coleslaw.

WINTER GREENS

If you have a layer of mulch on the ground, try tucking a young brassica plant, such as a cabbage or a cauliflower, through the mulch in summer and hope that it will grow away and come to fruition on a dark winter's day. It may succeed, but the level of soil nutrition is guesswork. On the other hand, if you are working the garden on a rotation, the winter brassicas should be planted after the early potatoes have been dug out in July. The seed potatoes will have had well-rotted manure or compost added to the soil to feed them. The potatoes will take up most of the nutrients in the manure but leave just enough to satisfy the brassica crop that follows. Remember, brassicas will not thrive in freshly manured ground.

Cabbages Depending on the varieties you choose to grow, cabbages will see

you right through the winter. Cabbages and kales are the most straightforward of the winter brassicas to grow, having little in the way of specialist requirements, needing only a reasonably fertile soil. They are easily raised from seed in either pots or outside in seedbeds. They germinate well and when they reach between 10–15cm (4–6in) high, they are ready to transplant outdoors. They also take up the least room in the growing bed of the brassicas – needing a distance of only 30cm (12in) between plants. Throughout Europe and North America, late summer is the last time for sowing winter cabbage. Once planted out, you can pretty much just leave the cabbages to grow. They will survive most conditions to produce a nice fat head. In terms of flavour, the crinkly-leaved savoy cabbages are beyond compare and far superior in taste and and crunchy texture to smooth-leaved varieties, while red cabbages have a flavour quite unlike other brassicas. Spring greens is a term used for winter cabbages that are planted out in the autumn and not harvested until the following spring. They need looking after throughout the winter by covering the soil with mulch and by checking the leaves for slugs and snails, but their wonderful taste is worth the effort.

Cauliflowers By far the most tricky of the brassicas, cauliflowers have a liking for two particular 'trace elements', boron and molybdenum. Soil with a high level of fertility will be rich in all nutrients and should be able to provide the cauliflower with what it needs. However, if you are gardening on very sandy or chalky soil, you are likely to be deficient in such trace elements, and there is no point trying to grow cauliflowers. They like to be planted in firm soil, so walk the ground before you plant, applying pressure from your heels. Cauliflowers will give of their best only if growth is unchecked. Long periods of drought will affect their growth unless they are watered regularly or planted in a deep mulch where moisture is retained. Cauliflowers are easy to raise from seed and, like all members of the cabbage family, will grow on happily as young plants until the time comes to reach maturity. Then problems can arise, and with cauliflowers it may be a poor formation of the head, discolouration or some other malformity that makes the head imperfect and inedible.

Brussels sprouts These brassicas produce tiny immature cabbages on the side of a tall stem and are perhaps one of my favourite. A surfeit of nitrogen may cause the sprouts to 'blow', which means that they do not set 'hard' but

BRASSICA CROP YIELDS

I have this nagging thought running through my head, saying we all eat far too much and unless we have large quantities on our plate we are not satisfied. In gardening, we must not fall into this trap of oversupply – whatever we manage to reap from a healthy plant is enough. Brassicas make us aware of this because some plants can give poor yields and you may find yourself picking only a handful of sprouts or a very small head of cauliflower. The chances are that unless we are experienced, full-time gardeners we are unlikely to build the levels of soil fertility that will give us the super yields of our dreams. Accept that our efforts have yielded what is right for us and our gardens – that is 'sufficient'. It must be stressed that gardening takes time and much attention to detail, it is hard work and often frustrating, and mostly it does not pan out the way the book says it should. We must not be disillusioned by this. In fact, we should feel safe in the knowledge that our best efforts will impact on a deeper level however many Brussels sprouts we did or did not reap from the single plant that grew well.

produce loose leaves. This does not happen if Brussel sprouts are planted after potatoes in the crop rotation, as the ground has not had a recent heavy application of nitrogen-rich manure. Sow Brussel sprouts in early springs and transplant the young plants to their final growing position by midsummer. By germinating a few different varieties that take different lengths of time to mature, you can have sprouts from late autumn through to the following spring. Then, at the end of the season, the top can be eaten and the plant will start to produce succulent side shoots (on which the flowers will be borne), so don't be too hasty in pulling up the plants. It is good not to always be in a rush to plant the next crop. Generally, the slower you are in the garden, the better you will be rewarded. Nature never does anything in a hurry.

Sprouting broccoli In terms of delicacy of flavour, this is the best winter brassica by far. Although it is easy to grow, the drawback is that it takes a year to mature, and is not ready to harvest until the following spring. The commercial sector has tried to hijack this wonderful vegetable by producing quick-growing varieties that flower in autumn. Do not be taken in by these; their flavour is a pale imitation of the real thing. Even though it is an annual, we grow it as a biennial. For homegrown purple or white sprouting broccoli, sow the seed in spring and transplant out the young plants by the middle of summer. They will grow happily in most soils without too much trouble, but the plants get big and might need staking if the garden is exposed and windy. They can take all the weather the winter can throw at them and still come up smiling. By early spring, you will notice side shoots growing at the leaf axils – the place where leaf meets stem. These side shoots are the beginnings of the immature flower buds. These flower buds are ready to harvest when the flowers have thrust out of their leaf case and produced a stunning purple head (or white in the case of white sprouting broccoli) the length of an index finger with a crown of purple. Properly steamed and coated with butter, this is as fine a winter vegetable as exists.

Kale Following on from the sprouting broccolis, there is a particular kale, 'Pentland Brig', which sprouts in spring and is eaten in a similar fashion. It grows in the same way and is quite excellent. The leaf kales, commonly known as curly kales, do not hold such charm. They are highly nutritious, but I find the flavour disappointing. Also, contrary to popular belief, they are not as frost-hardy as many people suggest, and a continuous bouts of heavy frosts will

Right: Baby turnips: one of the easiest and tastiest members of the brassica family.

How to cook Brassicas

Of all the brassicas, Brussels sprouts arouse strong emotions in people, and many hate them with a passion. I think the problem stems from sprouts being cooked incorrectly, as does our dislike for other members of this cabbage family. Overcooked, they turn bitter and smell foul; undercooked, they are inedible. Cooked properly, the flavour comes out well, and with these plants that flavour is subtle. Less so with cabbage perhaps, although savoys are in a class of their own, but with sprouts and broccolis the taste arrives when the vegetable is cooked through. Curiously, cooking through is almost to the point of being overcooked, but not quite. To get to this point and hold the flavour, the vegetables should be cooked rapidly in boiling water or steamed thoroughly, once the steamer is very hot.

Preserving cabbage

Long before deep freezes and aluminium cans were invented, food was preserved through the process of 'lacto-fermentation' – or pickling. In Japan, pickling is still very popular and no meal is complete without some condiment in the form of a pickled vegetable. Vegetables are the usual choice because they pickle well, but perhaps the best example of fruit being treated in this way comes also from Japan, where plums are pickled and eaten with rice.

Writing in *Nourishing Traditions*, Sally Fallon, the champion of proper food, encourages the use of the natural preservatives to prevent food from spoiling. She promotes the use of lactic acid, a natural preservative that also allows for better digestion of the foods because of the presence of digestive enzymes in the bacteria.

In Europe, the principle lacto-fermented food is sauerkraut, which is made from cabbage, whey, salt and caraway seeds. It is simple enough to make, lasts for months and has considerable health benefits. Beloved by Germans, who enjoy frankfurter sausages accompanied by sauerkraut, this is an exceptional food and one of the best uses of a cabbage that has survived the winter. As sauerkraut, you can store it for a further year, and the longer you leave it, the better it will get. This is the hard bit, keeping your hands off it, but do try because it is worth the wait.

To make sauerkraut, finely chop a head of white cabbage, add two tablespoons of salt and pound it with a potato masher or pestle for about ten minutes to release the juices. Next, press the cabbage into a large glass kilner jar and, if you have not used salt, add whey to start the fermentation process. Make sure that the glass lid is tightly sealed and airproof. Lacto-fermenation is an anaerobic (without oxygen) process and will not work unless oxygen is excluded. Caraway seeds may be added and stirred in before the lid is closed.

The sauerkraut will be ready to eat in a fortnight, but the longer you leave it, the more mellow the taste becomes. It is meant to be eaten as a condiment, and this is undoubtedly its best use for it is very strong but at the same time mouth-watering.

The whey that is used to make it is also easily come by. It is simply the liquid that remains after straining milk that has gone sour or curdled. It is protein-rich and full of nutrients and digestive enzymes, and is most widely used in the manufacture of cottage cheese. If you are going to make sauerkraut with whey, it must come from unpasteurised milk to achieve the maximum health benefits. As Sally Fallon points out, this is the only milk that is any good for you and that our bodies can digest because of the presence of those vital digestive enzymes.

damage mature plants of most varieties. Kales are simple to grow and produce leaves rather than heads. Nonetheless, I would prefer to concentrate my efforts and growing space on the sprouting equivalent.

Turnips and swedes This group of rooting brassicas are also easy to grow but should be sown in situ. They do not like to be transplanted. Once they have germinated, thin out the seedlings to the appropriate spacings – 10cm (4in) for dwarf turnips, 20cm (8in) for larger main crop turnips and for swedes.

Calabrese Summer broccoli is a joy. It is not hardy, so it needs to be sown in spring for harvest in late summer and early autumn. Sow seed indoors in late winter and plant out in early spring – the frost damages the mature head rather than the young plants, so planting out in cold weather is not the problem. Once you have harvested the main 'tree' of calabrese, lots of little side shoots will follow. There is also an ornamental variety known as 'Romanesco', which produces a head of extraordinary spirals. It is similarly tender but equally delicious.

Kohl rabi This is a stem brassica that forms a ball above ground with leaves on top. It is similar to a turnip in appearance and can be eaten raw as well as cooked. It has a sweet flavour, which is slightly more complex than that of the turnip but good nonetheless. They are best grown in summer because, like calabrese, the mature plants are not winter hardy.

Radish These come in all shapes and sizes from the small cylindrical 'French Breakfast' to the giant white mooli roots favoured in India and Japan. These are perhaps the first vegetables we learn to grow as children since they are quick and easy. All you have to do is sow the seed, water and wait.

DIRECT SOWING IN SEED BEDS

The traditional method of raising brassica plants was to grow them in outdoor seedbeds in open ground, for later transplanting. Unlike root crops (carrots and parsnips), young brassicas do not mind being transplanted and they are easy to germinate outdoors. (This goes for all brassicas except the rooting types – turnips, radish, Kohl rabi). Choose a sunny site in the garden for the seed bed – don't worry about where they fit into the crop rotation because the plants will be there only a few weeks. Clear away the mulch on the soil surface, if need be, and expose the bare soil. Make a seed drill using the tip of a planting trowel. Move the trowel back and forth in the soil to create a furrow with fine soil to either side. Sow brassica seeds at 3cm (1¼in) intervals. Cover over the seeds with your hand and firm down the soil gently. Water and mark the row with a label, stating the name of the seed and the date. There is no need for any thinning out; simply lift the plants and transplant them when they are 15–20cm (6–8in) high. Be delicate and try not to damage the roots as you lift the plants. Because of the deep tap root and the thin fibrous feeder roots common to brassica plants, most of the soil will fall off as you lift. Don't worry about this, but put each plant into a bucket of water so that the roots do not have a chance to dry out while they are waiting to be transplanted. Try and do your transplanting on a dull rather than a sunny day to prevent the young plants from drying out too much and to give them the best start in their new site. Select young plants that look healthy and strong. Leave weaklings behind and any that come up 'blind' (a deformed growing point that is easily recognisable). But don't be in a hurry to remove any remaining plants in the seedbed to the compost heap. You might have a disaster with some of your transplants and you may need spares. After removing plants from the seed bed, settle the soil by watering it.

PESTS AND DISEASES

Transplanting is the most vulnerable time in the life of a young brassica plant, and it is at this stage that you have to start checking them regularly for signs of pests and disease.

Cabbage root fly This merciless pest can trouble all brassicas, but turnips may escape attention. It lays its eggs at the base of young brassica plants, only for the grubs to hatch and feed on the roots, killing the plant. The plants droop and die. To combat cabbage root fly, place a 10cm x 10cm (4in x 4in) square of cardboard around the base of the plant, snug to the stem (cut a slit through the centre of the patch so that it can slide up the stem) and then cover with soil. It is a very effective method of protection.

Slugs and snails These will climb brassica plants and eat out the centre, so they must be dealt with. Check the plants on a daily basis and pick off any slugs that you see. Be especially vigilant in wet weather and come out at night with a torch, a couple of hours after dark, and pick them off the young plants.

Above: Hygiene and order are important in the propagation department; so, too, clear labelling.

Birds Pigeons are inordinately fond of brassicas and will be the scourge of the cabbage patch unless you deter them. They are cunning and brazen and will sit in a tall tree close to your garden and wait for you to go in at the end of the day and then come and get to work on your young plants. They will do untold damage, so you must keep them off. The best way is to net the patch against this threat when the plants are small. As the plants grow, the threat decreases and by autumn, when the pigeons are feeding on loose corn left over from the harvest in the fields and the brassica plants have become slightly more leathery and have lost the flavour of the first flush of youth, the problem will abate.

Cabbage white butterflies and moths Perhaps the worst problem of all are the larvae of two butterflies and a moth. The cabbage white, small white and cabbage moth. The winged insects are very noticeable through the summer and their progeny are soon hard at work, eating the brassica plants. The caterpillars of the large white have green and black-striped markings, while the small white produces green caterpillars, which are perhaps the deadliest since they burrow into the heart of the growing plant and eat it from the inside out, spoiling the plant as they go by destroying the growing tip. As with slugs and snails, check regularly for eggs and caterpillars. Look for the eggs on the under-side of leaves, where they start life, and also in the heart of the plants. They work quickly and you must be regular and thorough in your checking. Remove the affected leaves with the offenders to the compost heap by hand, where they will die having been parted from their food source. Alternatively, give the plant a good shake, which will release many of the eggs and some slugs and snails as well that you have missed. I also squash the eggs between my fingertips to destroy them. These methods are particularly effective in the autumn in wet weather, when both pests are active.

Flea beetle This jumping flea beetle can be a problem if allowed to feed on the small leaves of seedling turnips and salad rocket. Rows of seedlings should be covered with a protective layer of horticultural fleece or fine mesh netting.

Aphids These sap-sucking insects sometimes appear on mature brassica plants and need to be squashed, rubbed off or sprayed with a weak solution of an ecological detergent.

BIOLOGICAL CONTROLS

Increasingly favoured by organic farmers and growers, biological controls work on the practice of using a natural predator to kill a pest or a disease. They are very effective, but as with all state-of-the art innovations they are expensive. Available under various trade names within the industry, biological controls can be purchased in quantities small enough for the home-grower to use in their garden. There are many different types appropriate for a wide variety of crops – both field- and glasshouse-grown – but for brassicas there are two biological controls that will help out no end. The first is the slug nematode. This is a parasitic eelworm that lays its eggs inside the slug, killing it in the process. The worms are delivered by post, suspended in small packets of clay, which is dissolved in a bucket of water and then watered on to the soil. These nematodes are hugely effective and whilst they may not kill every slug in the garden (which is not what you want anyway, in the interests of the balance of nature) they will drastically reduce the slug population, particularly of the keeled slug, a small creature that actually lives in the soil rather than in dark corners of the garden, and which is much more plentiful and difficult to trace. The second control is a bacteria, which kills the larvae of the cabbage white butterfly and its associated species that feed on brassica plants. This bacteria, *Bacillus thirungiensis*, is sprayed onto the plants, especially the underside of the leaves, where the eggs are laid, and it destroys them and the larvae. It all sounds like the most horrific type of germ warfare and indeed it is, but what we must remember are two things – firstly you will get these pests thanks to the field scale growing of brassicas under chemical regimes and the immunity to pesticides that the butterflies have developed; and secondly both the nematode and the bacteria are naturally present and we are merely increasing their populations, which is why this form of pest control is accepted under organic certification regulations.

ROOTS (*APIACEAE*)

Baby carrots are what vegetable gardens were made for. Just ask Peter Rabbit. There are some crops that taste earthy – chard is one – but there is no finer indicator of a healthy and fertile soil than a newly pulled carrot eaten straight from the ground. It is one of those great gardening moments that cannot really be put into words. Whilst a summer carrot is an undoubted delicacy, I am not a great advocate of the winter root crops that hold up in the ground into the dark months. Winter carrots are certainly a good source of nutrition, but they take up a lot of room in the growing bed and, after harvesting, this ground is laid bare until the following spring. In terms of winter cropping, I think brassicas offer better value than carrots because once leaves have been harvested, more tasty young green leaves are produced. Most of the root crops, however, are relatively straightforward to grow, the only major problem being carrot fly, which can spoil that crop. For those with the smallest of growing spaces, roots are best avoided and – with the exception of baby summer carrots – roots are not suited to growing in containers.

In the ideal rotation system, the root crops occupy the ground that was held by the potatoes and the winter greens in the previous year. By planting them after these vegetables, there will be just the right amount of nutrition left in the soil to sustain them. (What they do not want are fresh applications of nitrogen-rich manure because it makes the roots split, or 'fang'.) By following an annual crop rotation, you will also help the roots defend themselves against carrot fly. The grubs of the carrot root fly pupate in the soil over winter, which offers a strong incentive to move the crop elsewhere the following year and make way for the legumes (beans and peas). The legumes will put back the nitrogen that was taken up by the potatoes, brassicas and then the roots, and the carrot fly will find its food source removed.

One of the great advantages of growing root crops for winter is that they store well and in the best possible place – the ground. Potatoes, onions, pumpkins and beans all have to be brought inside and take up valuable space, but winter greens, carrots, parsnips and even those root crops that produce leaves, such as chard and perpetual spinach, can stay outside. This is good news for the grower

Right: Celery grown in the old fashioned manner. Tying up with brown paper to keep the light out through the growing season produces a blanched, sweetened heart.

because digging up root vegetables, cleaning and sorting them and storing them in dry sand in a shed is an onerous task, a process to be avoided if at all possible. Much better to concentrate one's efforts on picking and storing crops like French beans. These beans can be picked on a sunny day, hulled in the warmth of the kitchen and dried on newspaper before finding their way into glass airtight storage jars.

CARROTS

Quick-maturing summer carrots are not in the ground long enough to come to the attention of the carrot fly, but the autumn and winter crops take longer to grow and will suffer accordingly. There are effective ways to combat the pest, one of which is erecting a barrier of fine netting around the crop. A crop cover, such as horticultural fleece, is also effective if you can keep it in place (difficult in a wild, windy autumn), but you have to ask yourself whether you are happy for a large portion of the garden to be covered in a white shroud throughout the growing season. There are biological controls available, but these are very expensive. I would grow something else instead of winter carrots and look to trade other vegetables for carrots, or just simply buy them in.

SOWING AND GROWING

If you are unmoved by this condemnation of carrots, they are easy to grow when sown direct into the ground, as long as the soil is warm. Sow the seeds in straight lines and 3–4 seeds per cm into a 1cm (⅓in) deep furrow in rows 15–22½cm (6–9in) apart. Thin out the seedlings when they are large enough to handle. Do this job as late as possible in the evening when the carrot fly is less active (it finds the carrots by smell, detecting the bruised leaves). Carrots will grow in any soil and will not require watering unless the summer is exceptionally dry. Sow more seed every six weeks.

PARSNIPS

These are very hardy and able to tolerate long periods of sub-zero temperatures. This means that you can have a fresh root available to harvest straight from the soil right through the winter, which is unusual in the cool temperate zone. Like carrots, parsnips should be sown in warm soil but at a distance of 1cm (⅓in) apart. Thin out the seedlings to 8–10cm (3–4in) apart. Parsnips are relatively trouble-free, suffering only from some skin canker, which is largely superficial

This is the term for the intermediate stage in the life of a plant that has been raised indoors before it is planted out. It is a process of acclimatisation and readiness for the outside world. Some members of the *Apiaceae* family, notably celery, celeriac and parsley, are very sensitive to changes in temperature, so when it is time to plant them out you need to introduce them very slowly. Start by bringing the plants out on a warm sunny afternoon for a couple of days, then for a whole day, then for a night (as long as there is no danger of frost) and then leave them out for several nights before you plant them.

Right: Despite an overly nitrogen-rich soil, causing excess roots to be produced, the parsnip grows straight and true.

and can be peeled off when they are harvested and come into the kitchen. One thing to avoid is sowing parsnips too early in spring. Firstly, germination will probably be sporadic because that is a trait of the family, but more importantly, if they do germinate, you will have giant parsnips by the autumn, which will be impossible to dig up and woody, fibrous and short on flavour. In the past, the high sugar content of mature parsnips made them a good crop to grow as fodder for livestock, but young tender parsnips are best for human consumption.

CELERY AND CELERIAC

Stem celery is absolutely critical, not for the crunchy stalks but for the white leaves at the heart, for this is where the true flavour is held. It is not hardy and a moody grower, so treat it with respect. Celeriac, which forms a bulbous root for eating, will respond to tender handling too. The problems arise when either celery or celeriac are planted out too early when it is still cold. This will lead them to sulk, stop growing and then bolt (run to flower and seed). Both celery and celeriac must be sown, 'pricked out' and raised in gentle heat 12°C (53°F), and when they are ready for planting out they must be 'hardened off' gently but properly (see left, 'Hardening Off'), as they are very sensitive to temperature variation. To ensure success, make sure the weather is warm before planting them into their final growing position in late spring/early summer. Planting distances should be 22½cm (9in) for celeriac and 30cm (1ft) for celery. Both celeriac and celery have a long growing season and they will not be ready to harvest until late summer and early autumn.

OTHER MEMBERS OF THE *APIACEAE* FAMILY

Parsley – perhaps the most widely used of all fresh culinary herbs – is also a member of the *Apiaceae* family, as is the rooting parsley 'Hamburg', which has white parsnip-like roots that can be stored in the ground. It was widely grown in the nineteenth century as the base for white sauces served with boiled ham. Hamburg parsley roots have exceptional flavour and are easily grown by sowing directly into the ground in seed drills and thinned to 10cm (4in) between plants. Leaf parsley, whether curly or flat-leaved, may also be sown outside in mid-spring when the soil has warmed up to at least 7°C (45°F). Sow it thinly in rows 1cm (⅓in) between seeds, but do not thin the plants to save time and keep the leaves small and tender. It is reasonably hardy and will continue to provide fresh green leaves well into the winter, although it will not survive consistently

HOW DO YOU KNOW WHEN SOIL IS WARM?

A soil thermometer will measure whether the ground is warm enough for planting (8–10°C/ 46.4°F is about right), but as you get to know your site you will be able to judge whether the ground is ready for planting seeds. There are days in spring when as you work outside you start to notice subtle changes in the soil: the smell of the earth is different, the sun dries the earth more quickly and the very topmost layer becomes warm to the touch. These changes might not happen until late spring, but it is worth waiting for this moment because it is when the ground is ready to receive seeds, particularly of the *Apiaceae* family, which are the most cussed of germinators.

hard frosts. Parsley also grows well in containers and can be sown inside in pots or seed trays for windowsill production.

OTHER ROOT CROPS

Other vegetables that can be grown alongside root crops, include beetroot, perpetual spinach and the chards. They conveniently fit into the root crop rotation, as they have similar needs in terms of soil nutrition. Like carrots and parsnips, they produce large roots, even though spinach and chard are grown for their leaf. They are all-important additions to the root bed because they provide fresh ingredients for cooking late into autumn and winter.

PERPETUAL SPINACH AND CHARD

These are grown primarily as biennial crops in that they are sown in spring, grow through the summer and autumn seasons and are left in the ground over winter, after which they flower and run to seed in spring. The reason why they are such important crops is that after recovering from a gruelling winter they put on a considerable amount of leaf growth before going to seed. They provide a valuable supply of fresh green leaves in the 'hungry gap' at the start of the year when there are few other greens of such high nutritional value available. For many growers, these two crops are a good investment. They are very resilient and continue to produce month after month without any problems. They have a reasonably high commercial value and crop for nearly twelve months of the year. Both are willing growers and are untroubled by pests and diseases apart from slugs and snails, whose impact is minimal. The secret of success with both perpetual spinach and chard is to let the leaves grow to a reasonable size before you harvest them. If you harvest the small leaves, the plant will never develop fully to reward you. This is important because it is the difference between a mediocre and a good crop and one that will be stronger going into winter.

SOWING AND GROWING

For bulk (and both spinach and chard reduce down to almost nothing when cooked), it is best to sow the seeds in 1cm (⅓in) deep furrows and crop them in rows 30cm (1ft) apart. So much is written about how there are no straight lines in nature, but when you want to raise large quantities of a crop and make it easy to maintain and harvest, growing in straight lines in a raised bed is a practical solution. One thing to remember with what appears to be the seed of chard and perpetual spinach is that the little crinkly cases are, in fact, fruits with two to four seeds inside. As such, you can sow thinly in the knowledge that from one seed case two or three plants will more than likely appear. Thin out to at least 10cm (4in) between plants and always water the row back in after thinning out (this applies to any row that you thin out of an open-ground crop or seedlings in pots). The soil and the roots must be re-anchored and the best way to achieve this is to water them.

BEETROOT

Like the parsnip, the beetroot does not improve with age. It becomes woody and loses flavour. To my mind, beetroot are best grown to golf-ball size and harvested as a summer crop when they are sweet and juicy. There are, however, varieties that are designed to hold their flavour and consistency over winter, but somehow they lack appeal. The difference between a baby summer beetroot and a winter soldier is marked. Beetroot grow easily when sown in short rows. There is no need to thin out the seedlings to produce golf-ball size beetroot, and they have no problems with pests and diseases. An added bonus is that the young leaves are edible and have a good strong earthy taste, so when you harvest a beet you have both root and leaf to consume. One thing to remember is that beetroot bleed, so it is best not to cut the root until they have been cooked. Cook them with their 'tails' on and remove the foliage by twisting rather than cutting.

ENDIVE (BELGIAN CHICORY)

Growing this crop successfully is like coming of age in the garden. It is a discipline that carries enough kudos to propel you into the upper tier of home-growers. When you produce beautiful heads of crisp yellow and white chicory at Christmas, you will astound your friends, and both your and their taste buds. It is a little time-consuming at the lifting and potting-up stage, but once that is over you can sit back and wait for the startling results. Belgian or Witloof chicory is worth all the parsnips in a winter, added to which the foliage makes excellent chicken forage or compost.

In spring, the crop is sown in rows 30cm (1ft) apart. Seeds should be 1cm (⅓in) apart and thinned to 10–15cm (4–6in) when large enough to handle. The conical lettuce-like plant is allowed to grow through the summer until it has reached maturity towards the end of the growing season in late autumn and before the first hard frost. The next stage is to lift the cylindrical parsnip-like roots and trim off the foliage to the neck of the root. Compost or feed the greenery to poultry and knock the excess soil off the roots. Only thick single roots are of use; those that have forked or are undeveloped should be composted. Three thick single roots are then stood upright in a 30cm (1ft) wide pot. Use soil, compost or sand to fill the pot. Allow the root crown to stand 3cm (1½in) above the soil level while still sitting 3cm (1½in) below the rim of the pot. Fit another pot over the top of the root crown to cut out the light. Transfer the covered pot to a warm greenhouse or potting shed.

FORCING, BLANCHING OR BOTH?

The aim of the exercise is to replace the lush green foliage of summer with pale underdeveloped leaves. This is achieved by growing the roots in the dark and warm without access to light. It is also known as 'forcing' the crop because by bringing the potted roots inside, you will encourage what would otherwise be a dormant root to grow out of its season. It takes about six weeks from potting up to harvest, depending on how warm your greenhouse or potting shed. The ideal temperature is anywhere between 5–12°C (41–53°F). To liven up the salad bowl, there are red and white variegated varieties of chicory that blanch a beautiful pink. You get only one chance at this, once the roots have produced the 'chicons', they are spent and should be composted.

SALSIFY AND SCORZONERA

Both of these winter root crops are coming back into fashion, particularly with foodies. Salsify has a thin white tapering roots like a small parsnip but a less scented and more sophisticated taste. Scorzonera grows straight and has rough black skin. Both have a somewhat smokey taste and it is this that makes them stand out. They are delicious roasted and break the monotony that accompanies winter root crops. Salsify and scorzonera have the same growing requirements as parsnips, so grow best in soil light with organic matter. They are sown thinly into furrows in open ground and thinned to roughly 10cm (4in) between plants. You should allow about 30cm (1ft) between rows.

In summary, this group of root crops are forgiving. They grow without much fuss in almost any soil. Sandy soil is easier for root vegetables to penetrate so, if you are gardening on heavy clay, add some sand, gravel or compost to break it up a bit. Be vigilant for carrot fly, even if you have covered the crop with net or fleece, because this is a most determined pest which renders the crop inedible.

Right: Belgian chicory may produce one large chicon or many small ones; all equally delicious and a rare winter salad treat.

LEGUMES

DIRECT SOWING OR POT RAISING?

Leguminous crops, such as peas and beans, have big seeds, which makes them easy to handle and sow direct. Both vegetables also grow well in containers, although you will need to know how tall the variety you have chosen will grow so that you can provide support. All can be sown directly into the soil into their final growing space, but the seeds and seedlings are much fancied by slugs and snails and mice. Watch out for slugs and snails, particularly in a wet season. To avoid slug, snail and mice attack, some growers like to raise seed under cover in pots for planting out at a later date, to increase their chances of success. A pot-raised plant will shoot away quickly in warm soil, whereas a direct-sown seed is at risk from pest attack the moment it germinates. For the 'sufficient' grower, however, any vegetable that can be direct sown should be, to save time and space.

Peas and beans are important in the productive vegetable garden, as they play a vital part in returning precious nitrogen to the soil, so helping build soil fertility. These legumes take atmospheric nitrogen and hold it in bacteria that are found in nodules on the roots of the plants. (When you dig up a bean or a pea plant, you can clearly see the pink wartlike nodules on the roots.) This ability to take an invisible gas from the air and turn it into soluble plant food is truly extraordinary. That the fruits of the plants are highly nutritious, protein-rich and palatable to both humans and animals is an added benefit.

If we include legumes in our crop rotation, the addition of nitrogen to the soil will be a natural occurrence without the need for wheeling in heavy barrows of manure. The legumes fit into the rotation plan after the roots but before the onions, shallots, garlic and leeks (alliums). After having grown root vegetables, the ground will still be very fertile but not too nutrient-rich, which legumes prefer. (Freshly manured ground will only encourage lush leaf growth at the expense of the pods.) And once the crop is finished, cut off the tops of the plants and leave the roots with their nodules to return the nitrogen to the soil, ready for the alliums to use.

PEAS

There is nothing quite so delicious as homegrown, freshly podded peas from the garden. I like to toss the raw peas into a salad bowl, as they taste so sweet and have a wonderful crunchy texture – it seems almost a shame to cook them when they are this fresh. But peas are an 'in and out' crop and are best off grown early in the season. Later on, just as blight gets hold of the potatoes, the pea moth, thrips and mildews will find the peas and yields are often reduced. To ensure sturdy plants, choose early varieties of peas that do not grow high off the ground, and disregard varieties that claim to be self-supporting; they will not be.

SOWING AND GROWING

Because of the size of pea seed, there is no need to create wide, shallow drills for planting. It is just as easy to dib individual holes in the soil (or through the mulch) and pop in single seeds, 1cm (⅓in) apart and 2cm (⅔in) deep. Rows should be about 60cm (23½in) apart. As the pea plants begin to grow, they will produce tendrils to help them climb. The tendrils will need something to cling to, so drive some twigs into the ground or use a length of netting to support the plants if you are growing peas in rows. Pea plants produce quite low yields when you consider the amount of time, effort and space they occupy in the vegetable garden. However, peas are delicious and, being legumes, they will benefit the soil. A better option yield-wise will almost certainly be the mangetout varieties or sugar snaps, but compared to a fresh garden pea they are somewhat pedestrian in flavour.

BEANS

So many types and varieties, so much easier to grow and much less fussy than peas, beans are a staple and an absolute banker for the home grower. They like sun and warmth, have no particular soil requirements and, when sown in succession, will feed you with fresh beans through the summer and dried beans through winter. You can make baked beans, salt runner beans or successfully freeze any type of green bean. They will add nitrogen to the soil and encourage bees into the garden in great quantities. The trick is to know when the beans are young and tender and at their best for eating and to know when to take out the crop to make way for something else, such as a 'catch crop' of lettuce, which can make use of the precious nitrogen that the beans have left behind in the soil before winter sets in. Many types of bean are climbing plants, so they can be grown in a limited space as long as there is room for them to grow upwards – this is a hugely useful virtue where space is at a premium. Also, most beans can be dried for later use in soups or stews, or kept for seed to sow the following year.

BROAD BEANS

One major attribute of broad beans is that some hardy varieties can be sown in open ground in autumn and will crop at least two weeks earlier than conventional spring-sown broad beans. There are varieties to take you right through the summer, culminating in the Windsor strain from where we get the very British, Brown Windsor Soup, but really the joy of broad beans lies in the first and juiciest crops of early summer. Autumn-sown varieties will germinate

soon after sowing, reach 10cm (4in) in height and then stop growing almost completely over winter until they take off again in early spring. The young plants will not be unduly troubled by the cold or any pests and diseases, so it is well worth considering an autumn sowing. This early cropping has obvious commercial advantages, but also gives the home grower time to clear the ground and plant out a late summer or autumn crop of salad leaves, such as radicchio, which will enjoy the leftover nitrogen.

SOWING AND GROWING

Like peas, broad bean plants need support, as they have no climbing mechanism. The traditional way of growing broad beans in double rows works best. A double row consists of two lines of beans about 20cm (8in) apart with 5cm (2in) between each bean seed. Each seed should be dibbed in 2cm (¾in) deep. The next set of two lines should be 60cm (24in) away. The double row system ensures that the plants can lean into one another for support during and after wind and rain, and aids pollination. However, a double-row planting will not prevent the plants weighed down by bean pods falling to one side. To prevent this, drive stakes into the ground and wrap string around the block of beans.

Few pest or diseases will harm the crop. You may get brown markings on the leaves, which is caused by a fungus called 'chocolate spot'. This indicates a lack of potassium in the soil, but it is only superficial. Likewise, the pea and bean weevil may chew the leaf edges, but nor will this harm the crop.

Try to eat broad beans when they are young and tender, when their flavour is at its best. When the bean crop is harvested, cut off the stalks just above ground level and remove to the compost heap. Cut up the stalks into small pieces, otherwise they will not rot down properly because the air spaces will be too great between them and the bacteria and fungi that do the work in the compost heap cannot come into play. Leave the roots that hold the nitrogen-filled nodules in the ground to feed the soil until you have to take them out to plant the next crop. Save and dry some of the broad bean harvest for next year's seed.

RUNNER BEANS

These traditional British beans were once a firm favourite of the vegetable gardener. Runner beans are a classic example of plants that produce far more beans than we can possibly want to eat. Rather than planting out a long row of beans, we probably need only two or three plants maximum. (There are also

Right: Beans come in all shapes and sizes and legumes are essential for returning precious nitrogen to the soil.

dwarf varieties that can be grown in pots, which may be a better option.) This is an important lesson for the 'sufficient' grower – not to waste too much time, energy and space on growing plants that produce surplus.

Runner beans need canes or string to grow up and plenty of sun. Once they have germinated and the plants have started to grow, they are foolproof and will climb away happily on their own. In dry weather, they like to be well watered, as they have a rather feeble root system. Like broad beans, runner beans are best when picked and eaten young. When you have finished harvesting the beans, take out and chop up the plant foliage for the compost heap. Dry and save some of the beans inside the pods for next year's seed.

FRENCH BEANS

There are some exceptional beans in this group with excellent flavour, including the fat striped borlotti beans and string-thin flageolets. Some French beans are extracted from the pods and dried while others are eaten as green beans. This difference in type will be clearly indicated in your seed catalogue. Also indicated will be whether they are climbing (pole) beans or dwarf (bush) beans. Climbers are like runner beans in that they need supporting canes but, once growing strongly, they will cling to the canes on their own. Do not to plant too early. A cold spring is their worst enemy, and soil temperatures need to be in excess of 10°C (50°F). Rain cools soil rapidly and increases slug activity and slugs favour French beans. If a wet spring is the pattern, raise French beans (both climbers and dwarf) in pots, and plant them out when they are growing strongly.

The art of drying French beans is straightforward. Leave the beans on the vine until they become brown and papery. Then the beans should be harvested and hulled before being left on a flat surface in a light airy and dry room for a week or so before being stored in airtight glass jars. Do not apply heat or direct sunshine, just leave them to dry naturally in a dry atmosphere. They will store over several winters, but should be eaten within twelve months, except for those required for seed next year.

How to cook beans

BROAD BEANS

Choose small pods, as the beans will be young and have a better flavour than bigger, older beans. Very young broad beans, less then 7cm (3 in) long, can be cooked in their pods and eaten whole. Older beans must be podded and are best skinned to remove the outer coat which toughens with age. The easiest way to do this is to slip the beans out of their skins after blanching. Allow about 250 g (9 oz) of whole beans in pods per person. Cook in boiling salted water for 8–10 minutes or until tender. Skin if necessary. Serve topped with melted butter and chopped herbs. Older beans are ideal for puréeing or making into soups.

FRENCH BEANS

Choose young tender beans. Allow about 125 g (4 oz) per person. To prepare, cut off the ends. Most varieties are stringless but if not remove the strings from the seams of the pods. Cook in boiling salted water or steam for 5–7 minutes until tender. Serve hot or cold – these are delicious with a simple vinaigrette.

RUNNER BEANS

Runner beans become stringy with age, so are at their best when young. Choose beans that break with a crisp snap. Allow about 125 g (4 oz) per person. To prepare, cut off the ends and remove the strings from the sides. Cut into fine long slices, or cut young beans into short lengths. Cook in boiling salted water for about 10 minutes, depending on age, or until tender. Serve with melted butter.

DRIED BEANS

The weight of dried beans approximately doubles during soaking and cooking, so if a recipe calls for a certain amount of cooked beans, start with half the amount of dried beans. Soak overnight in plenty of cold water and then drain off the water prior to cooking. Add salt only 10 minutes before the end of cooking times; if added at the start it will toughen the skins of dried beans.

ALLIUMS

Members of the onion family, including onions, shallots, leeks, garlic and even small salad onions, follow on in the crop rotation from legumes, as they grow well in the nitrogen left behind in the soil by the beans. As many home-cooked meals begin with an onion, it makes sense to grow some of your own, although they are cheap to buy. As with maincrop potatoes, you might feel the space and time is better taken up with something less easily available. However, unlike potatoes, onions are not troubled by sinister diseases such as blight, and are easy to grow given plenty of sunshine and a normal season's worth of rain.

ONIONS

The easiest and quickest way to grow an onion is from a 'set', which is simply an immature onion that you plant directly into the soil in early spring. Very few varieties of onion, however, are available as sets and many more varieties are available as seed. But growing onions from seed is time-consuming and takes up space. In my allium patch, I tend to grow a few onions from seed, some winter leeks – because they are easy and take absolutely no time or effort once planted – lots of spring onions through summer and autumn, which I grow from seed, and a few shallots from sets because they are have a wonderful flavour, and certain sauces, such as bearnaise sauce, cannot manage without them. I do without growing garlic, as it is prone to disease that can lead to crop failure and would much prefer to be growing in a Mediterranean climate.

GROWING FROM SEED AND SETS

To plant onion and shallot sets, first rake the soil to a relatively fine tilth (breadcrumb size is ideal). Then gently push the sets into the soil up to the neck, leaving the little twist of dry skin sticking out at the top, although it will not matter if they are pushed in a little deeper. Plant the sets 10cm (4in) apart. This is all you have to do until harvest, except keep the onion patch hoed and free of weeds. Onions do not like competition from weeds, and as monocotyledons (single-leaved plants), they need the nutrition in the soil around them for themselves. Also the weeds create humidity around the base of the onion

Below: Almost a staple in the kitchen, the decision is whether to grow or buy in.

plants, which can lead to fungal problems. You will get onions if the patch is left to become weedy, but they will be much smaller for cooking, and small onions do not store so well.

Onion seed can be sown in late winter in gentle heat indoors. Plug trays with twenty or more modules are excellent for the job. Sow two seeds to each module and, if two germinate, remove one, but only when the seedlings are growing strongly. Grow them on until the seedlings are about 10cm (4in) high and the roots have started to move around the base of the plug. After 'hardening off', young plants can be planted out. The slender young onion leaves look fragile, but they are tough and hardy – only the most violent of rainstorms will knock them about.

SHALLOTS

It was always said of shallots that they should be planted on the shortest day and lifted on the longest day. Like onions, shallots are tough and can be planted any time during winter. In practice, it is best to plant them in late winter or as soon as it is dry enough to get on the growing bed and rake over the soil to make it suitable for planting. Plant each shallot bulb in the ground up to its neck, about 8–10cm (3–4in) apart, and from this a clump of bulbs will grow. At the end of the growing season, their skin will start to crisp up and they will be ready to be harvested and brought inside to dry off fully. The best way to store both shallots and onions is to plait strings, like the Frenchman with the bicycle, and hang them in a cool, well-ventilated place. They keep well over winter. Like garlic, but not onions, shallots can be saved as seed to plant the following season.

SPRING ONIONS

These little salad onions are grown from seed and should be sown in short rows, as a little goes a long way. They are invaluable in the kitchen and make a good substitute for an onion early in the year before onions are ready late in the summer. They require only a fine tilth in a seedbed sited in the sunshine. The seed should be almost but not quite touching in the row. They will grow away slowly, but once they are mature, sow another pinch of seed in an adjacent row 15cm (6in) away. It will pay to do this 'successional sowing' from spring right through until autumn for a continuous crop of fresh, young spring onions.

LEEKS

Along with turnips, the French call leeks 'les rois du potage', the kings of soup. They are absolutely essential for any productive vegetable garden. Leeks require a minimum of skilled cultivation but impart wonderful flavour, especially at the intermediate stage of their life. The great old leek foot soldiers that are still upstanding in the garden at the end of winter will be very woody and lack flavour, but if you start leeks in spring and sow them successionally, you will have a crop of tender young leeks from summer until late winter.

SOWING AND GROWING

For early season crops, sow leek seed in a seed tray or a fruit box lined with newspaper in early spring. They do not need heat to germinate but prefer some cover, such as a cold frame, an unheated greenhouse, or a polytunnel. When the leeks are about 10cm (4in) tall, they are ready to plant outside. For a winter crop of leeks, sow the seed outside direct into a seedbed. When the plants reach 20cm (8in) high, lift and carefully transplant them to their final station. Allow 10cm (4in) between plants and 20–25cm (10–12in) between rows. Before transplanting the young leeks to the final site, 'top and tail' each one to reduce stress on the plants and give them a better start in life. To do this, take a pair of secateurs or a sharp knife and cut off one-third of the foliage and one-third of the roots before dropping the plant into a 10cm (4in) deep hole. Water in each plant by filling the holes with water. Gradually, over the next few weeks, the holes will fill with soil and the shank of the leek will be blanched in the darkness as it grows.

Left: Room should always be made for quick-growing spring onions.

CUCURBITS

The *Cucurbitaceae* family, which includes pumpkins, squashes, cucumbers and courgettes among others in its ranks, has become increasingly popular in recent years. The joy of pumpkins and squashes is that they can be left alone to grow in a corner of the garden, as long as it is sunny, while the cucumbers and the courgettes need their fruit picked regularly. Otherwise, they will produce inedible giants suitable only for chicken feed or the compost heap.

CUCUMBERS

Cucumbers can be split up into various different groups. Greenhouse varieties need protection and support and are a little time-consuming because the side shoots need pinching out and the main stem tying in. Ridge cucumbers will grow happily outside in a sunny bed and produce masses of fruits. There are also numerous different baby varieties for pickling, which are variations on the ridge cucumber. Ridge cucumbers will climb as well as sprawl across the soil but, if you are worried about slugs, encourage them to grow up a support. The only require-ment for these crops is plenty of well-rotted manure. Compost is adequate but, like potatoes, cucumbers are hungry, so when you plant them, dig a proper hole a 'spit' deep (the depth of the spade head), and put two spade-fulls of manure in the hole. After planting, put a mulch of manure on top. Use well-rotted manure and water the plant well before you lay the mulch. This goes for any mulching – the ground beneath must be damp first; otherwise the rainwater has to get through the mulch before it can add any moisture to the soil.

GLASSHOUSE CUCUMBERS

If you insist on a 'proper' cucumber, smooth, long, dark green and ribbed, then it is the glasshouse cucumber for you. This is perhaps the only time I will suggest an F1 hybrid seed. The reason is that the F1 varieties available are 'all female' hybrids and are therefore guaranteed to be sweet. They should be grown as cordons (single-stemmed plants) up a string or a cane and be planted with plenty of well-rotted manure like their outdoor cousins. The key to success is to pinch out the leaf buds at every node while leaving the tiny cucumbers

Right: Keep the growing space well-ventilated and damp, for pest- and disease-control, to ensure a bumper crop of cucumbers.

'The joy of pumpkins and squashes is that they can be left alone in a corner of the garden to get on with it, as long as it is sunny.'

untouched. Damp down the floor of the glasshouse with water twice a day to deter red spider mite and keep the beds well mulched. Feed weekly with comfrey or liquid seaweed. Cucumbers will also do well in containers.

COURGETTES

Courgettes are so giving: they keep on and on producing their fruit and deliciously edible flowers right through the summer and autumn. To keep the plant producing courgettes, you must pick them when they are small; otherwise they will grow into marrows. Once the first marrow forms, there is no way back to baby courgettes. (The plant now thinks it is time to concentrate on seed production and so will jump quickly from baby to adult fruit.) To grow well, courgettes require sunshine and a good dollop of manure at planting. Allow 60–90cm (23½–35½in) between plants. Water them well once and leave them. They cover some ground in their sprawling manner, but they are not a trailing plant like most pumpkins and squashes. When you sow courgette seeds in pots, place each seed on its edge in the potting compost. If you sow them flat, water may gather on the seed and rot it. Be sure to 'harden off' the plants properly before planting out, and wait for the weather to warm up because none of this family is frost-hardy.

PUMPKINS AND SQUASHES

As popular as they have become, pumpkins and squashes take up a huge amount of space for a relatively small return. It takes a lot of leaf on a pumpkin plant to produce a couple of fat orange specimens for Hallowe'en (which seems to be the reasoning behind a lot of pumpkin growing), and I would rather use my sunny corner for courgettes. Having said that, there are some wonderful varieties available, and they will grow quite happily on a rough piece of land that you don't quite know what to do with. Simply dig a decent hole and get some manure into it, plant a pumpkin, water it and forget about until the autumn, and you will be nicely surprised. For me, this is not a crop to spend time on. All this family will germinate readily on a sunny windowsill, and given manure, space, sunshine and water at planting will grow away happily. They are an edge crop and do not fit into any particular rotation, untroubled as they are by any particular pest or disease.

Below: Pumpkins and squashes are becoming increasingly popular, with a huge varietal choice.

PERENNIAL VEGETABLES

For low impact, no-dig, minimal interference growing, we have to take a long look at perennial vegetables, as they present more opportunities for hands-off gardening and low soil interference than growing annual vegetables. Once they are planted, that is it: no further tilling of the soil is required. The key vegetables, such as asparagus and globe artichokes, are special and have significant monetary value, which makes them all the more worthwhile growing. There is only one golden rule, and this applies to anything perennial planted anywhere in the garden: all perennial weeds must be removed. Bill Mollison, one of the founders of the Permaculture movement and a great advocate of perennial crops, was quoted as saying that if you have to do something on the site, you really don't want to do, make sure that you do it well once so you never have to do it again. It is good advice, and never more appropriate than when planting perennials. You cannot dig up the plant once it is in the ground, so you must get the perennial weeds out first, and that means digging over the area very thoroughly before you plant.

ASPARAGUS

What a luxury this crop is and – as with all perennials – it is easy to grow once established. However, like all good things, you do not get very much and not that often. For a bed to yield a decent amount of asparagus, it needs to be at least 4m (13ft) long by 1m (3ft) wide with plants grown in double rows 30cm (12in) apart and 30cm (12in) between rows. The plants will produce spears of asparagus between spring and midsummer. After this period, stop cutting the asparagus spears. This will allow time for the ferns that protrude from the spears to grow and return sustenance to the crown beneath the ground. By the end of the year, the fern will turn brown. Cut it to ground level and mulch the entire bed with well-rotted manure or compost.

You can sow asparagus seed in pots and plant them out the following spring. Alternatively, you can buy one-or two-year-old crowns ready for planting, which will be quicker to crop. Dig out a trench to a 'spit' depth (the length of the spade's head) and plant the asparagus 30cm (12in) apart. Replace the trench soil.

Below: Of all the perennial vegetables, asparagus is the most luxurious.

Then mound up the soil by throwing up soil from either side of the trench. This will aid drainage and help to blanch and tenderise the spears. Asparagus should not be harvested until their third year, and then only with restraint.

GLOBE ARTICHOKES

These grow very well from seed and quickly. In spring, when the soil has warmed up, they can even be direct sown into the ground, though you must watch out for slugs, as they enjoy the young leaves. They like full sunshine and plenty of water at planting because they have a poor root system for the amount of foliage they have to support. The plants form beautiful clumps of jagged foliage, from which the globes are produced in midsummer. By the second season, you may well get some artichokes. Instead of harvesting all of the globes for the kitchen, leave some to develop into flowers late in the summer. The flowers are a wonderful blue colour and make excellent forage for bees and insects. Once established, artichoke plants need little or no attention, except perhaps some thinning out every few years, as the plants are vigorous and can become overcrowded. Globe artichoke plants will die down in winter in the coldest areas, but are hardy and will survive most conditions. If you are in an area that has regular hard frosts, give them a good mulch of straw or bracken to protect them. To keep the plants productive, you can renew the oldest plants by division. Take off the side shoots from an existing plant along with its underground buds or suckers – these produce the new growth, roots and foliage. Plant the rooted suckers at least 1m (3ft) apart. Keep them well-watered until they are established.

Cardoons are like a giant form of artichoke and are used more for ornamental rather than culinary purposes. Victorian gardeners wrapped the side shoots of cardoons in hessian sacking and blanched then in spring for eating in late summer. This is well worth doing if you have the time and inclination. Not quite as delicate as sea kale but good nonetheless as an alternative to braised celery or fennel.

SEA KALE

A brassica native to the pebble beaches of the south coast of England, sea kale is blanched undercover of pots (dustbins or buckets) as it starts to grow from its crown in spring, to produce white and very succulent leaves. The taste is sweet and mildly cabbagy when steamed, but the texture is crispy, juicy and an excellent vehicle for melted butter, just like asparagus. Quite a delicacy, sea kale is very easy to grow and has a considerable 'wow' factor. In spring, sow seed in pots. Plant out in early summer and grow on outside as for any large brassica (if you can obtain root cuttings known as 'thongs', so much the better because they are quicker to establish and produce good strong plants). Come winter, the foliage will drop off and the plant die back to two or three dormant buds on the soil surface. Mulch these and then leave them to grow again for another season. In early spring of the third year, cover the young shoots with a pot and keep the plant in total darkness until the white shoots are ready for harvesting. Replace the pot after harvesting and you will get another flush three weeks later. After that, remove the lid and let the next crop of leaves grow in the light and turn green. This will replenish the plant for next year's blanched crop.

JERUSALEM ARTICHOKES

Nothing to do with Jerusalem or artichokes, these tall plants are rooting sunflowers that produce a lot of foliage and an excellent edible tuber, which can be stored like potatoes. They are annual plants, but I have put them in the perennial section because they stay growing in the same place year on year. This is because they are impossible to move on account of the fact that they will produce more plants from any single piece of plant with a bud attached that is left behind in the ground. Pig or chickens will clear the ground of Jerusalem artichokes, but really they are best left to grow in the same site. No matter how thoroughly you dig the ground each autumn and winter, you will always leave some tubers behind for the crop to regenerate fully the following year. This way they can be counted as perennial vegetables.

RHUBARB

Rhubarb is unfairly thought of as an old-fashioned, rather one-dimensional crop but, packed with vitamin A and producing delicious edible stems very early in the spring when almost no other fruit is available, it is invaluable for the home grower. The plant has virtually no known enemies apart from slugs, which are attracted to the succulent pink stems, but as the weather and soil temperatures are still cold when the plant is ready for harvest, the slugs are not that active and do little damage.

As a perennial whose leaves die off in winter and one which can last for ten years or more while requiring little or no attention, rhubarb is a must for

Right: Globe artichokes are also widely grown for their succulent hearts and decorative qualities.

the garden. The plant's huge decorative leaves also add an ornamental feel to a neglected corner, where it will grow in either full sunshine or even a little shade.

BLANCHING OR FORCING?

The first act of human intervention comes towards the middle of February, when the dormant crown of the plant, from where the leaves will protrude, should be covered with a large bucket, pot or even dustbin. As the weather begins to warm up and the plant returns to growth, the new leaves and stems will start to appear from the prominent buds on the crown and begin to lengthen. Because they are growing in the dark and are unable to photosynthesise properly, the leaves will be yellow and malformed and the stems a translucent pink (as opposed to red if they were growing in full daylight).

This act has wrongly become known as 'forcing' when it is in fact 'blanching'. Strictly speaking, the term 'forcing' applies to crops that are forced out of season, something that was a great feature of Victorian horticulture when owners of large gardens desired specific crops at odd times of the year.

After two or three croppings (the stem parts easily from the crown with a twist and a pull), remove the cover to let the plant grow in full light. Large leaves will develop quickly and help to reinvigorate the plant by taking energy back down into the crown. Further cropping of the by-now red stalks can continue into early summer, but the plant should be allowed to grow uninterrupted for the remainder of the summer and autumn.

After the foliage has died off, the crown should be thickly mulched with well-rotted manure or compost, and the mulch, under which the plant should always grow, should be topped up.

Rhubarb may be grown from seed or bought as young plants from nurseries, but probably the best way to start propagation is to find a gardener with a rhubarb patch and ask for a division in winter. A sharp spade driven into the crown will produce a new plant as long as there are dormant buds on the top and some roots attached to the yellow stem.

Right: Rhubarb is a valuable source of vitamins early in the season. Terracotta forcing jars, stuffed with straw to keep out the light help to produce the very best.

growing vegetables

SALADS

The gigantic mix of leaf crops, herbs and wild plants which is collectively known as salad is perhaps the most important nutritionally of any group of vegetables. The moment a section of a plant is removed from its parent – a leaf is cut, a fruit picked – it begins to lose nutrition because its food source is removed. It continues to lose nutrition when it is cooked. If we can eat vegetables in their most vital state, which is raw or as fresh as possible, we will get the best out of them. This is not a plea on behalf of raw food as a diet, it is merely a suggestion that 'fresh is best' and that salad finds plants at their most nutritious.

For the first time in the vegetable section, we start to get away from the rigidity of cultivated plants and the formality of families and rotations. Certainly there are some plants in this group that need to be rotated around the garden but salad, as a generic term, brings in all kinds of things – weeds and wild plants as well as more commonly known cultivated salad crops, such as iceberg lettuce – and a homegrown iceberg is one of the finest lettuces of all.

With wild salad leaves and edible flowers, we have a connection to the wider garden and hedgerow, and as we come to know the soil and native flora, we can begin to embrace the concept of 'sufficient' in more real terms. It is to be in touch with what is around us and to learn how to work with it, accepting that it is very bountiful and giving. 'Sufficient' hangs on this very theme – that we have so much, if only we open our eyes and look. I will expand more on this theme in the Wild Food section (see pages 188–201), but it starts in the garden, where it is important to appreciate that there are fresh edible ingredients beyond cultivated rows of specific vegetables.

LETTUCE

However much we plan to incorporate wild plants into our gardens, the bulk of the salad harvest, especially if you like to eat a lot of salad, will still have to come from cultivated plants.

The bedrock of salad, for me, is lettuce, partly through conditioning and partly because it is very simple to grow and good to eat. The range of lettuce is huge, from butterheads, crispheads, upright cos types, looseleaf, oak leaf and

Left: Despite the arrival of countless varieties of salad 'leaves', lettuce remains the number one for me.

baby leaf lettuce mixtures, all the way through the season to winter lettuce varieties and bitter endives. All lettuces have similar growing requirements: they are not fussy about soil type, they like full sun, the young leaves are eaten by slugs, but the plants are pretty tough. In summer, growth may slow down or the leaves may bolt in hot, dry conditions. Propagating lettuce from seed is simple. You can raise lettuce in a seedbed outdoors from late spring onwards and then transplant them into permanent beds. Alternatively, you can sow seed into 9cm (3½in) pots for pricking out into seed trays, planting them out when each one has a small rosette of leaves. When planting out, make sure not to bury the crown (to avoid rot) and water them in. At night, keep a close eye out for slugs, as they will eat their way through the tender young leaves. Be conservative when sowing seed; lettuce, like brassica seeds, are reliable germinators and you do not want a glut of lettuce all at once. Instead, sow a small amount of lettuce seed every three weeks to maintain a regular and constant supply throughout the summer. By mid-to late summer, it is time to sow the endives and chicories that prefer cooler temperatures and, in autumn, turn your attention to those lettuces that can survive into the winter months, whether planted inside or out.

Lettuce are more than happy to grow in containers. The baby lettuce leaf mixes that provide a selection of colours and types grow particularly well. However, avoid seed mixes that include rocket and lettuce together because the rocket will grow much faster than the lettuce and upset the balance. Successional sowings of these seed mixes will keep you in fresh salad leaves throughout summer and autumn.

THE ROCKET PHENOMENON

Brassicas have come to the forefront of our 'salad-leaf thinking' in recent years, with rocket the champion. A hot, peppery leaf, rocket comes in two forms – salad rocket and wild rocket – and has taken the salad world by storm. It is not as hot as some of the Chinese brassica leaves, such as the mustards, but the flavour of rocket is exceptional and it is small wonder that it so popular. Others to consider are the crinkly mizuna, giant red mustard, and the various different forms and varieties of pak choi. The leaves are the edible parts, but the flowers of most of these brassica leaves are edible too.

Salad brassicas, however, are not immune to some of the threats of pests and disease that trouble the larger leaf brassicas, such as cabbages. They are badly troubled by flea beetle at the seedling stage and pigeons can take an

Below and Right: Many salad leaves are in the brassica family and are best eaten at the baby leaf stage.

interest too, as can slugs. They are much safer from these dangers when grown in containers rather than open ground. When brassicas are grown in open ground, they can fit in anywhere in the rotation as long as they are not repeatedly grown in the same spot. Their nutritional requirements are minimal – owing to the fact that they grow to only a small size – thus barely robbing the ground of nutrients. Like lettuces, salad brassicas need full sun and will grow well when sown thinly in rows. A covering of fleece will deter the flea beetle.

CUT AND COME AGAIN LETTUCE

The concept of 'cut and come again', whether in containers or in open ground, suits the growing pattern of baby lettuce and the salad brassica leaves well. The seedlings grow through to true leaf stage and are then cut down to just above the crown of the plant so they can regenerate. There are two benefits: the first is that you get leaves at their most tender and the second is more than one crop.

SPINACH

I have talked of perpetual spinach (spinach beet) in the root section (see pages 122–23) but annual spinach, the delicious velvet-textured green leaf plant, is different altogether. It has an annoying tendency to run to seed very quickly and thus is best grown as a baby leaf plant. Also, the leaves shrink down when they are cooked, so you need a great amount to make a dish. Eaten raw as salad, it is both delicious and nutritious. Sow it in rows in the garden or direct into containers, but do not pot raise and transplant; this will not work. Space the seeds 1cm (⅓in) apart and apply 'cut and come again' rules.

GOLDEN PURSLANE

A personal favourite, this plant has thick leaves and stems, which are succulent and tasty. It has an exotic feel to it but is easily grown in the open ground in rows. The seed is tiny. Sow it thinly to produce small round leaves.

CORN SALAD

Lamb's lettuce, or mache (as it is known to the French), is a slow-growing but very hardy low-growing salad plant that produces a single rosette no bigger than the palm of the hand. It is very useful for winter production and has a delicious velvet texture, like cooked annual spinach. Sow the seeds thinly in rows throughout the year; it grows as well in winter as it does in summer.

HERBS

The choice of fresh herbs available to buy is very poor compared to what we can easily grow in our gardens or even on our windowsills. Fresh or dried, herbs are needed only in small quantities yet have a big say on flavouring food. In terms of 'sufficient' that snip of fresh chives can make the difference between feeling empowered enough to want to grow plants rather than leave it to someone else. Herbs have inspired many to start a growing career. And remember that whatever we do in the way of growing plants will bring about change, no matter how small. It is an honouring of nature and the elements, even if it is something as simple as growing a clump of chives in a terracotta pot. Chives are a good plant for the 'sufficient' grower and are a permaculturist's dream because they have many benefits. Chives are perennial, hardy, drought-tolerant, edible, have edible flowers, attract some beneficial insects and ward off others, look and smell beautiful, and can be divided up and moved any-where in the garden, including the shade. The more perennial herbs that we can grow, the better. Once we have them, they are there to stay. Some perennials, such as rosemary, marjoram, sage and tarragon, will need cutting back and, because of their Mediterranean origins, prefer well-drained soil and full sun. Others are best grown as annuals, including coriander, parsley and basil, and need some care in propagation and planting out. Most herbs are well suited to pot culture, and aside from those which are frost-tender, such as basil, most can be grown outside all year without problems.

Culinary herbs are best broken down into three groups for growing purposes: annuals, perennials and self-seeders.

ANNUALS

Parsley Perhaps the most famous of all, parsley is adaptable to both pot and garden soil but as with all members of its family, *Apiaceae*, it likes warm soil to aid germination. It can be sown in seed trays in the kitchen, though it germinates more successfully in a heated propagator, but if you want to sow parsley directly outside in open ground, leave it until the soil has warmed up to 10°C (50°F). Parsley is in fact a biennial plant (it flowers in its second year) but we grow it as an annual to produce the best tender edible leaves, which come in many forms and varieties from the standard 'moss curled' to giant French or Italian flat-leaved.

Basil This easy-to-germinate annual will grow to twice the size if it is kept undercover rather than planted out. It likes heat and light and rich soil, with plenty of compost rather than manure.

Coriander Plagued by running to seed in the twinkling of an eye, this pungent Asian herb is best sown in situ later in the season. For some reason, spring sowings bolt; perhaps it is the fluctuations in temperatures. Wait until early summer and accept coriander as a herb for harvesting in summer and autumn rather than spring. It can also be grown for its seeds, which can be dried naturally on the plant and collected to use in cooking.

Chervil There are a number of herbs that have the flavour of aniseed, including the fennels, but by far the most delicate is chervil. This easy-going annual can tolerate some shade and its fresh aniseedy leaves liven up a salad.

Fennel There are three options here: an annual leaf fennel; an annual bulb fennel, known as Florence fennel (whose white bulbous stem we braise or chop raw into salads); or a perennial leaf fennel, which dies down each winter. Straightforward all: just add plenty of compost and allow full sun. Space the Florence fennel bulbs at 10cm (4in) apart; you will need only one perennial bronze leaf, and the annual can grow in a row with seeds 1cm (½in) apart thinned to 3-4cm (1¼in).

PERENNIALS

Sage This herb needs full sun but not too much in the way of nutrition. Herbs of Mediterranean origin prefer a weaker soil, so you do not need to add any compost or manure to the soil. Buy plants from a nursery or garden centre and plant them out in full sun. To avoid woody stems, trim back sage plants by a third each year in spring.

Rosemary This beautiful plant also gets woody and, like sage, needs cutting back, but only after flowering in summer or autumn.

Thyme Low-growing, often prostrate in habit, thymes also need full sun and no added compost or manure. They too are best cut back after flowering and before winter sets in.

Marjoram An essential herb, known as oregano in the US, marjoram is a strong and low-growing perennial, which needs nothing more than the dead flowers removing in autumn. It is less fussy about nutrition than thyme and sage and can be easily grown in rich soil.

Sorrel A pinch of sorrel seed sown outside in a row will bring forth delicious lemon-flavoured leaves throughout the summer and autumn for several years. This hardy perennial should be grown and cropped in a strong fertile soil like perennial spinach or chard, but is not quite as vigorous.

Tarragon French tarragon is the one to grow because it has exceptional flavour unlike Russian tarragon. The plant is not fully hardy and will die down and disappear in winter. It may need a covering of mulch to keep off heavy frost. It likes the sun and needs some compost at planting.

Mint The point to remember with all the varieties of mint is that they spread. Roots run underground and shoots pop up everywhere. They will become tangled up with the roots of other plants and are generally a menace, so mint plants need to be grown in one large container on their own, where they will thrive for years. Feed mint in spring and midsummer with a liquid fertiliser to keep the plant vibrant, and cut it down in autumn to allow for some small fresh leaf growth over winter.

SELF SEEDERS

Dill One of my favourite herbs, dill is grown as an annual that will self-seed. The seedlings may overwinter in mild areas or the seed germinate the following spring, thus providing a plentiful supply. The seedlings may be easily lifted and transplanted.

Borage With pretty blue edible flowers, borage is a perennial with good insect-attracting qualities (particularly bees). It self-seeds prolifically and may need to be thinned out because the plants grow large and prickly, like comfrey.

Chives The bedrock of the herb garden, the chive plant is perennial but will self-seed, thereby giving you the means of propagation. The plants are also easily divided into clumps in the cooler seasons of the year.

Right: Chives are one of the most valuable, multi-functional herbs in the garden.

FRUIT

GROWING FRUIT

Strawberries apart, with most fruit we enter the realms of trees, shrubs and climbers. Different sets of skills are required for the disciplines of fruit growing: some pruning, perhaps some training and tying in, and most of all patience. Watching and aiding the development of a woody fruit plant over the seasons and years is a very rewarding experience, and has the making of us as gardeners and growers. Most vegetables are relatively straightforward in their growing, fruit less so. A lettuce may be ready to harvest in 45 days of planting the seed, whereas an apple tree may continue to produce abundantly for 45 years. With annual vegetables, we can change our tastes and try out different varieties each year on account of their short growing period. With the apple tree, we are in for the long haul; there is a sense of permanence about it as a long term investment. With the skills mastered, the return on that investment is huge.

Growing perennial plants perhaps has more of the feel of 'sufficient' about it because the process is a slower one and moves away from the rough-and-tumble of non-stop gardening that surrounds the production of annual vegetables. This is never more evident than in the homestead gardens of Asia, where huge quantities of food are produced on tiny plots of land, around a dwelling. Usually the only food that is bought in is the staple, rice, and the condiments, in the form of salt, sugar and oil for cooking. Forest gardening, which is becoming more popular in the developed world for this very permanence, allows fruit trees and shrubs and, to some extent, annual plants to grow together in the same way as the Asian homestead gardens.

If you only ever do one thing in the garden, make sure you plant a tree. It is a signal of permanence, a doffing of the cap to the future, an expression of intent.

Trees give us many things: they stabilise the soil, prevent erosion, encourage soil life, give shade and shelter, amusement and timber and last, but perhaps most importantly of all, they give us the oxygen we breathe. That they produce fruit to nourish us is but one of their qualities. If we look upon them as harbingers of all these important things and see them as much more than simply fruit producers we will establish a rapport with them, which will help both them and ourselves to thrive.

We are fortunate today that there are fruit trees which can thrive in small spaces and containers, as this reduces the requirement for large parcels of land. Nonetheless, if you are going to plant a tree in the open ground, you need to plan because it will not appreciate being dug up and moved. You need to know how big it is going to get; where it will cast its shade and whether this will inconvenience other plants; where the root system is going to spread; and how you will use the ripe fruit and manage the fallen leaves.

As to the fruit itself, there is a wealth that we can grow to suit all soils, climates and tastes. Obviously the cooking and storing potential of many fruits is enormous as well as the nutritional content when consumed fresh. There is one other thing: growing fruit well is something to be proud of because at times it can be very testing.

FRUIT TREES

This is the generic term for hardy tree fruit, the important ones being members of the *Rosaceae* family; apples, pears, plums, cherries, and the more exotic peaches and nectarines.

APPLES

With the banana, which is the fruit most consumed by humans on the planet, the apple has been domesticated for thousands of years. Growing apples is as basic as it comes, providing you remember some key points:

1. They are not self-fertile – that is, they need cross-pollination via pollen from a compatible variety of apple tree.
2. They need shelter for bees and other insects to do the pollinating work.
3. Apple varieties are grafted onto rootstocks, which control the vigour of the trees. You need to know what that rootstock is to determine how big the tree will be when mature.
4. Apples are geographically sensitive. Climate and, to a certain extent, soil will determine how well they perform.

CHOOSING VARIETIES

The fact that apples are 'geographically sensitive' is very important when choosing varieties. Establish which varieties grow well in your area, as this is a good guide. It is absolutely pointless ordering expensive trees only to find that they do not thrive. Examples of this abound. Generally, apples like high summer sunshine and low summer rainfall which, while not essential, are certainly preferable. Cox's Orange Pippin, one of the best known of all apple cultivars, likes these conditions. It will not tolerate poor light levels or high rainfall, which make the trees susceptible to fungal diseases and severely reduces the crop. Talk to gardeners, visit the local fruit farm or 'pick-your-own', and find out which apple trees produce good reliable fruit crops and which do not.

Right: The apple retains the top position for fruit growers in the temperate regions of the world.

PLANTING

Most of the basic rules for planting are the same for all trees. You only get to do it once, so do it properly. It makes no difference whether you are planting a tree in a field or garden or against a wall for training; they all need the same amount of care at planting. The size of the hole you dig depends on whether the tree you are planting has been raised in a container or in open ground. If the tree has been raised in open ground, it will arrive from the nursery with no soil around the roots (bare-rooted). Pot-raised trees have more of a rootball. Either way, the hole needs to be deep and wide enough to accept the root. Spread out the roots of a bare-rooted plant before half-filling the hole with a mix of compost and the soil itself. Then, taking the stem by the hand, jiggle it up and down a few times. This will allow soil particles to fall into the gaps beneath and amongst the roots. Firm the soil gently with your foot before filling in with the rest of the soil and firming the soil again. Then you should water. Don't bury the 'graft union' – the awkward join usually found 15cm (6in) above the root. If you do, it will encourage the rootstock to grow, which is not the idea. If you are in an exposed area where the tree might need staking – consistent wind rock causes a gap to appear between soil and stem, which allows water in, leading to a condition known in the trade as 'The Death' – you must drive the stake in before planting the tree to avoid root damage. After planting, prevent weed growth by placing cardboard or newspaper around the tree and putting well-rotted manure or compost on top of the cardboard. Young trees do not like competition from weeds. Most garden soils are acceptable and, left alone, apples should reward you. There are some fairly insignificant pest and disease problems, but healthy soil and a sunny site will see a happy apple tree. The most likely problem will come from the codling moth. To prevent the moth grub climbing, put a greaseband around the tree trunk. Also a fungus called scab may put black spots on leaves and fruit. The best way to control scab is to rake up the leaves once a week through the autumn and put them in a pile to make leaf mould. Any bacterial canker may be pruned out over winter.

SPECIALIST SHAPES AND PRUNING

Restricted forms of growth, such as espalier, cordon, fan-trained and even stepovers (single-tier espaliers), are greatly to be admired. They also have solid space-saving claims in the small garden, when grown against walls and fences and beside paths. Fruit in the *Rosaceae* family are all suitable for training. However, specialist shapes take time, and training fruit trees is a skill that takes some learning. It is perhaps best to learn this skill from a specialist book dedicated to pruning and training. I would favour a free-standing apple tree over a trained one any day. There is no greater pleasure than walking amongst the gnarled old apple trees that live out their days at the bottom of the garden.

As far as regular pruning goes, you can get away without pruning and still have lots of fruit, no matter what size rootstock the tree is growing on. A tree wants to grow and it wants to produce fruit. It does not need anyone hacking into it every year to hurry things up. By the time you have a mature tree, you will have more fruit than you know what to do with and, unless you are a veteran juicer or cider maker, you will be glad that the windfalls are eaten by thrushes and blackbirds.

RENOVATING OLD TREES

This is a common problem: how do you deal with the old, neglected tree that has been ignored for years? The answer is, in stages rather than all at once. Take out all topmost branches each year over the course of four or five years. What you can and must do immediately is to remove any dead or diseased wood or branches that are crossing and rubbing against one another.

Also of value to an old tree is to scrape away any grass and thatch around the base. Prick holes in the topsoil with a garden fork and give the area a thorough mulch of homemade compost. This will work its way into the soil via the fork holes and help to reinvigorate the tree.

PEARS

Pears can be recalcitrant. You think you are providing the ideal conditions for them and they will not respond. Then, when you do get fruit, it never seems to ripen. The pear-ripening problem is solved by storing the fruit at the right temperature; some varieties ripen at room temperature, others like it cooler. In order to provide the best growing and ripening conditions for pears, there are several points to bear in mind in order to achieve the best possible results.

Pears come into blossom earlier in the spring than apples. This puts their blossom at risk from frost. Next, pears are more delicate than apples. Hardy certainly, but less tolerant of cold winds and draughts. They need a warm niche in the garden, but they also dislike being crowded. They are at their best when

'When mature, the pears reach a certain size and fullness of shape and the fruit skin has a bloom.'

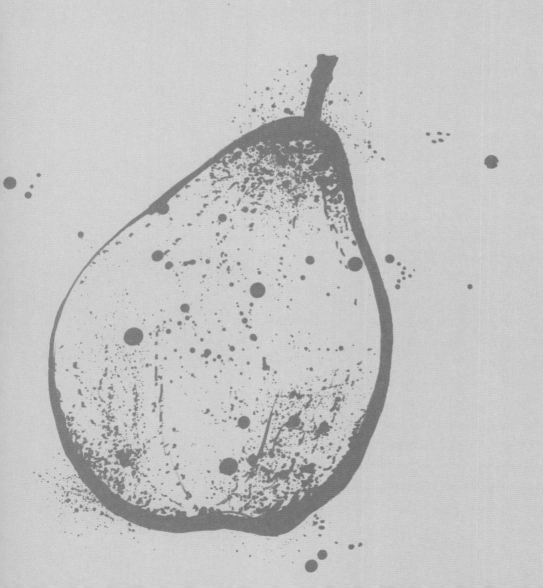

trained as a fan or espalier against a warm but not fully south-facing wall. This is what they love, the heat bouncing off the stone or the brick and being released late in the day when temperatures drop. Finally, pear trees do not need full sun all day; a little bit of shade is preferable.

In terms of soil needs, pears are not especially fussy and they need pruning only if they are trained against a wall on a system of horizontal wires. Free-standing pear trees can look after themselves. Unlike apple trees (grown on dwarfing rootstock), pears do not grow well in containers. They will resent the root restriction and fail to flourish.

STORING

The storage method you use depends upon the variety of pear you have grown. Conference pears, one of the best known of all dessert pears, do not ripen on the tree. When mature, the pears reach a certain size and fullness of shape and the fruit skin has a bloom. Conference pears must be picked in one session, or they will go over on the tree. Store the fruit in a cool frost-free place, preferably in the dark. When the fruit is required for eating, it should be brought up to room temperature (i.e. in the kitchen) for two weeks. This last bit is crucial. Even after two weeks the fruit might still feel hard to the touch, but the flesh yields to the bite and the pear will be fully ripe.

I think it is worth persevering with pears because a ripe 'Doyenne du Comice' pear is one of the finest fruits in the world. Pears are not as reliable and easy to grow as apples, but the fruit is so luxurious that if you get them right, they are well worth the effort.

PLUMS

Plums are a very good choice of fruit tree for a garden with a little bit of shelter and some reasonable soil. They suit the 'sufficient' grower well because they allow copious quantities of ripe fruit to be picked off the tree (one of the great pleasures of gardening and growing generally), and there are many ways plums can be stored as juices, jams or bottled whole. With the right cultivar and given the optimum conditions, plum trees can bear heavily, so much so that fruit-laden boughs of the famous dessert plum 'Victoria' have been known to snap under the weight of the fruit. As well as the damson, the group also includes the wonderful greengage, perhaps one of the finest stone fruits grown in temperate climates apart from peaches and nectarines.

It is a seminal moment when a fruit tree of your own planting begins to bear fruit. A plum tree can take five years or more to produce a decent crop, but it is well worth the wait. Because of this and the fact that any tree is an investment for life and maybe even beyond, it is critical to get the position of the plum tree right when you plant it. Apple trees are much less demanding than plums. Plum trees will not tolerate badly drained soil and low-lying cold areas, particularly frost pockets which lead to the added threat of the blossom suffering damage. Plums need the late summer sun to ripen. A south-facing wall or fence is perfect, but this much sought-after place in the garden should go to peaches if you intend to grow them. (Both peaches and plums grow well as freestanding trees, but peaches will do that little bit better when trained against the wall.)

Like pears, plums are early flowering. Fruit production can be severely hampered if frost gets to the flowers, so a warm, sheltered place in the garden is ideal. This early flowering means that windy, exposed sites are not good. The pollinating insects need warmth and calm conditions. The pollinating work is done mainly by honey bees, who know they are better off staying in the hive if the weather is cold, wet and windy.

The choice of rootstock is important for plums. These trees get big but on the dwarfing rootstock, such as 'Pixy', they will stay compact. If possible, resist the urge to prune plum trees, or the tree may contract the deadly silver leaf fungus (I have seen it kill entire trees). This is a fungus whose spores are active in the colder months of the year. It can enter the plant through a pruning cut, so if you must make any cuts to live tissue, do this in summer to autumn.

PEACHES

It is supremely important that we derive satisfaction and pleasure from the fruits of our labours. Yes, we need the staple foods, 'the plodders', such as the leeks and the lettuces, but we also need the luxuries like beautiful ripe peaches. It may surprise you to know that homegrown peaches are yours if you want them, and they are much more reliable to grow than pears or plums. Sounds too good to be true, but trees, especially fruit trees, want to flower and produce fruit. It is their job. Peaches flower as prolifically as cherries and plums because this is the nature of stone fruit – they produce masses of flowers in spring. Just look at the hedgerows in spring, and you will see the white blossom on blackthorn (sloe) – another member of the Prunus genus. Wild

Right: Plums can be heavy-bearing and reliable given optimum growing conditions.

plums and bullaces are the same, producing clouds of blossom. In the garden, it is our concern to try and protect that blossom and make sure that it turns into fruit.

A peach can be grown as a freestanding tree or trained against a wall or fence. The aspect must be south-facing, and the more shelter, the better: full sunshine for as much as the day as possible is essential for the fruit to set and ripen. Peaches flower early and although frost hardy, the blossom is not.

The peach is the tree to choose for the one precious south-facing wall in the garden – the prime spot. And believe me, when you produce bowls of tempting ripe peaches, you will encourage more people to take to the soil and start gardening than if you were to serve up any other fruit or vegetable.

PRUNING AND TRAINING

Free-standing apple and pear trees can manage without pruning, especially if they are grown on dwarfing rootstocks, but peaches grown in northern climes of the temperate zone do best if trained against a wall or fence. This affords them maximum heat and light while the background support will help to distribute any trapped heat when temperatures drop at night. With training comes the need for proper pruning to ensure that the peach branches and shoots are in the right position to benefit.

HOW TO TRAIN A FAN

The best shape to choose is a fan rather than an espalier because it allows the peach tree shoots to be tied in odd places and at odd angles. At planting, let the tree grow through the first year without any pruning. The following year shoots will start appearing at every bud. You need to rub out 50–75 per cent of the shoots, except for two or three near the base of last year's growth and the leader at the end of the branch. Allow the leader to grow to the length you want it and then pinch out the tip. All the fruit will appear between the beginning of last year's growth and the new season's leader growth. The following year, the two or three buds that you left to grow at the base of the original shoots will carry the fruit. You then have to go through the same process again of rubbing out the shoots as they appear and leaving the fruit behind. Each season, the old shoots are tied into the structure to create the fan.

Feed peach trees weekly from the moment the fruits set until the peaches start to change colour and ripen. Then stop. The feed should be a mixture

Below: Climate change is facilitating the production of peaches outdoors in the cool temperate zone.

of 50:50 liquid manure and liquid seaweed. Water well weekly, but split the timing between the liquid feed and watering; a three-day period is perfect. In very dry periods, hose the tree once a day with a fine spray of water to deter red spider mite, especially if the tree is growing indoors in a glasshouse, where it is more susceptible to attack.

PEACH LEAF CURL

As a word of warning, there is a fungus called peach leaf curl. Peaches grown outside will suffer badly from this fungus if they are not covered in winter. To protect the tree from attack, cover the entire tree from about 30cm (12in) off the ground with tough clear polythene or some other material that light can penetrate. A frame can be built around the tree and the polythene attached to the frame. Keep the tree covered through the winter until it comes into blossom in early spring. Then remove the cover to allow pollinating insects to reach the blossom – most peaches are self-fertile, but a little extra help never goes amiss.

FAMILY TREES

In his excellent book *Designing and Maintaining your Edible Landscape – Naturally*, Michael Kourik favours the use of genetic dwarf trees and family trees. Where space is at a premium, 'two-for-the price-of-one' thinking is a big advantage, and this is what you get with a family tree that has had different varieties of apple or pear grafted onto the rootstock. This means you can have early, mid-season and late-ripening varieties on the same tree. They may be compatible for purposes of pollination too. Genetic trees are dwarf by nature – not because of the rootstock they are grown on – and so are ideal for the small garden and for container growing. The thought of a tall, unmanageable tree that casts a lot of shade can worry many gardeners. Much better to have something that is appropriate for the space, and this is where dwarf genetic and family trees on dwarfing rootstocks can help.

SOFT FRUIT

The term soft fruit covers the bush fruit, primarily currants, gooseberries and their various hybrids. The main ones that are worth concentrating on are blackcurrants, which make the finest jam; redcurrants, from which we get a sumptuous jelly; and the whitecurrant, which is high in pectin and helps the first two set properly. Gooseberries are a dream, either eaten raw or cooked, but seem to be undervalued these days for whatever reason. I have always adored them and they are very easy to grow. Sucking the flesh out of lots of ripe gooseberries is one of the summer's great pleasures.

All these fruits grow as bushes, usually on a 'leg', which is a short single length of stem from which the main framework of branches protrude. The exception to this is the blackcurrant, which grows as a 'stool' with a number of shoots coming from the crown of the plant at ground level.

Tolerant of most soils and also a little shade, currants and gooseberries are low maintenance in every possible way except for pruning in winter. Once you get the hang of this, there is little else to worry about. With a little enterprise, the reds, whites and gooseberries can even be trained as cordons – single-stemmed plants growing to 1m (3ft) and higher – particularly useful if you have limited space. For most households, one plant of each is enough to produce a considerable yield. Soft fruit bushes are a good alternative to tree fruit (if you don't have the room or feel overwhelmed by the thought of trees) because they are litle trouble and yield a lot of fruit for small plants. They will all do well in containers, and for the initial cost of the plants they give a good return.

BLACKCURRANT

The primary difference between the currants and the cane fruit is that the best comes from currants when they are cooked. Certainly they may be eaten raw, but they come into their own when turned into jams, jellies, juices and sauces. They also benefit from the addition of a certain amount of sugar, for they are quite tart (and my cheeks are pricking at the very thought of a mouthful of raw redcurrants). Gooseberries are different; there are some dessert varieties which sweeten up very well when ripe and require no extra sugar.

Although the leaves of the different currants look similar, blackcurrants are easily told apart because, when crushed, the leaves give off the aroma of blackcurrant juice. None of the other currants have these fragrant leaves.

PLANTING AND GROWING

Chances are, you will buy a two- to three-year-old soft fruit plant from a nursery because that is how they are marketed. Alternatively, you can easily propagate your own, as they 'strike' well from hardwood cuttings. This can be done in autumn. Make sure that the soil is finely worked, and cut one-year-old wood from a bush (currants and gooseberries) and bury it to four buds deep, leaving 10cm (4in) above ground. By spring, the wood will have rooted and may be lifted and transplanted to its permanent site or potted up in a large container filled with a soil and compost mix.

Nursery-bought currants and gooseberries can be planted any time of year, although spring and autumn are always best because there is more moisture in the ground for the young roots to access. Shovel some well-rotted manure or compost into the planting hole (as you would for any shrub) and mulch around the surface to keep down annual weeds, and prevent moisture loss as well as adding nutrition. Make sure that all the perennial weeds around the planting hole have been removed because, as with trees, you will not get another chance to weed deeply in amongst the roots.

Redcurrants, white currants and gooseberries are pruned differently to blackcurrants, but all currants and gooseberries fruit on wood made the previous season. In the first year, just leave the plant to grow. Then in the winter of the year after you have planted it, remove 30 per cent of the older growth of the plant at ground level. With blackcurrants, that percentage of old growth needs removing every year. The reason for removing the wood at ground level on blackcurrants is that the new growth will come from buds that are still under the soil on the main stem of the plant. These buds will grow for a year, then fruit the next, and after a further year they may be removed to make way for a newer shoot.

My professional training in commercial horticulture taught me to prune currants and gooseberries each year. The rule book says that regular pruning will improve yield and allow light and air into the plant, which in turn will keep fungal diseases and pests at bay, as well as keeping a sensible size and shape of bush for ease of handling and picking. But that is typical of the rule

book. With currants, just as with fruit trees, you can be 'hands-off' and be a 'no-pruning grower' and still get plenty of fruit. However, this approach will not prolong the life of the bushes, which will quickly become oversized, straggly and unproductive.

So much of gardening is about intervention and, increasingly, many of us feel we would rather let nature take its course. There is an organic mindset that says we should 'live and let live' and not worry about low yields. Indeed, this has become fashionable thanks to the work of some highly acclaimed growers and writers, such as Masanobu Fukuoka (*The One Straw Revolution* and *The Natural Way of Farming*). With some plants, this 'hands-off' approach is fine; with others like currants and gooseberries, there may be a less happy ending.

REDCURRANTS AND WHITECURRANTS

The only difference between the two currants is the colour of the fruit, and I think no fruit garden is complete without either one or the other. There are few problems in the growing. They have no special requirements except half a bucket or a good shovel of compost at planting and a permanent mulch around the base of the plant to hold in moisture and suppress weeds. No extra feeding is required. I also prune them in winter to maintain a good yield of fruit.

The vibrant colour of the fruit are very attractive to birds. The real problem with this is that the birds that come to feed on them tend to be common garden species that are comfortable around humans. As such, they are not easily scared off. Purpose-built fruit cages are expensive and often unsightly but a covering of some description may be essential once the birds find the crop. If this is the case, then a net slung over the currant bushes should do the trick. It need only be in position from the time the fruit starts to ripen until it is picked. If you have only two or three bushes, it is not too much bother, especially considering it might make the difference between a bumper crop for you or the birds.

PRUNING

As with blackcurrants, allow the red and white currant plants to grow for a year without doing anything to them. Thereafter, pruning should be carried out in the dark months of winter, when the plant is dormant. Take the 'leader' (the foremost shoot, the tip of the branch) and cut it back by one-third and all the side shoots to three buds. Any shoots that are crossing over one

Left: Currants are versatile fruits with few problems and high levels of cropping guaranteed.

another or pointing from the outside into the centre of the shrub may also be removed. If you feel that the centre of the plant is becoming very constricted, you might want to thin it out to improve ventilation. However, such overcrowding is unlikely in the first few years.

GOOSEBERRIES

As far as pruning goes, gooseberries follow exactly the same programme as red- and whitecurrants. The only thing you might want to do as the gooseberry bush increases in years is to prune out low-lying branches that scrape the ground because they make weeding difficult. Also, gooseberries come into flower earlier in the season than the currants and so you must not leave your pruning late, otherwise the sap will be rising. There is less need to cover the bushes to protect the fruit from birds, but there is a pest called sawfly which can defoliate the bush in a day, so be vigilant. Sawfly lays its eggs early in spring, and there may be several hatchings. The stripy caterpillars have either to be shaken or picked off and removed. Try putting a bird feeder sock nearby to encourage the tits and robins to eat them.

One thing that gooseberries do appreciate is potash. This is most readily available in the form of wood ash from the house fire or bonfire. Potash is highly water-soluble and, as such, drains freely from the soil in a wet winter and is taken up by the plants. It will do the bushes all the good in the world to have a bucket of wood ash spread around them once or twice during winter. Cut comfrey leaves are another good source of potash, and these leaves can be spread around the surface of the gooseberries, which, as with currants, should be spaced at least 1½m (4ft) apart.

STRAWBERRIES

A bowl full of ripe strawberries on the kitchen table is well within the reach of most of us with reasonable garden soil and a little time to carry out important operations at critical times. Strawberries are not fussy about soil, climate or temperature. They are fully hardy, but they do like summer sun and only the small, near-wild strawberries 'fraises des bois' will grow in shade.

The simple fact is that strawberries are a cinch to grow but incredibly impressive; let's hope your success can encourage your friends and neighbours to start growing for themselves.

Right: Gooseberry varieties have declined over the years, but both culinary and dessert are still available.

THE PATH TO UNBLEMISHED FRUIT

The three main enemies of strawberries – and none will appear until the fruit is ripening – are birds, slugs and a fungal disease called botrytis (grey mould). To protect strawberries from birds, cover the crop with a fine mesh net. This will prevent birds from forcing their way into the net and becoming tangled.

An application of a biological control will severely reduce the slug population. Biological controls are expensive, but they do work. Otherwise, you must come out each night and pick the slugs off the strawberries and the surrounding crops. This method will damage the slug population, especially if you remove slugs early in the season before they have a chance to breed. Raising strawberries in pots and barrels off the ground is also a very good way to avoid the slug problem, and it will also prevent botrytis.

Botrytis is a grey mould fungus caused by soil splash onto the fruit and by humid conditions under the leaves and amongst tight trusses of fruit nestling between the soil and leaf cover. Crowding the plants will also raise humidity, so space them out in the beds at least 30–45cm (12–18in) apart. Commercial growers cover strawberry beds with strong black polythene and plant through it so that the fruit stays dry sitting on top of the polythene. For me, this use of polythene is not acceptable. I prefer the old-fashioned way of mulching around the plants with straw so the fruit can rest on the straw and remain dry. This is far more attractive and just as effective.

DIFFERENT STYLES OF GROWING

Hanging baskets are good for growing strawberries, as the fruit is able to hang down. Also the 'runners' – the tiny strawberry plants (roots and all) that appear from the centre of the plant on long stems and are the means by which strawberries propagate themselves – like to hang down over the side of the basket. In the garden, the runners sprawl over the soil and take root every 20cm (8in) or so. They can be snipped off, potted up into 9cm (3½in) pots and planted out, once the roots have grown enough to fill the pot.

Specially designed strawberry tubs and towers, where the plants are planted in through holes in the side of the container and from which the fruit hangs down as it does from the basket, are also excellent for preventing bird, slug and botrytis attack. They allow free air movement, which will combat the fungus, and the slugs will not be able to climb up the barrel before you spot them. Birds also find these containers difficult to deal with and are less brave if the

Right: The strawberry in both forms, wild and cultivated, is a high quality fruit worth the pains involved in the growing.

tub or tower is kept close to the house.

Growing strawberries in either hanging baskets or barrels is great, but nothing, absolutely nothing, beats the taste of strawberries grown in garden soil. All strawberry plants require are some compost at planting, but no heavy application of manure. Naturally, plants grown indoors will grow quicker and ripen faster, but humidity can be a problem undercover. Ventilate the fruit properly, especially on sunny days.

AFTERCARE

You should get three years of full cropping from a strawberry plant before it will start to lose vigour and need replacing. The method is simple. When the runners have produced small plants with noticeable roots, pot them up in homemade compost and keep them watered until the roots have filled the pot and it is time to plant them out. This may be done either in the autumn of that year or the following spring. Those runners that you do not need must be cut off at source to prevent the plant from being weakened and to prevent the strawberry patch from becoming a tangled mass of plants of different ages. Whether potting up or planting out in the garden, make sure that the growing point (crown) of the strawberry plant remains prominent above the soil level. If it is buried, there is every chance the crown will rot and the plant die.

CONSUMPTION

We are used to buying strawberries that have been refrigerated and wrapped in plastic, but to get the best out of homegrown strawberries eat them as soon as you can after picking. On no account put strawberries in the fridge. Hurry up and eat them, preferably as a meal on their own with a little cream and sugar.

'Strawberries are not fussy about soil, climate or temperature. They are fully hardy, but they do like summer sun and only the small, near-wild strawberries 'fraises des bois' will grow in shade.'

CANE FRUIT

Raspberries are the most important of the cane fruit, but there are also some very useful and vigorous berries that can be encouraged to sprawl over buildings or waste ground or up supports and fences. Some of these are hybrids, such as the loganberry and the tayberry, while others, such as the cultivated blackberry or the Japanese wineberry, are pure-bred, 'stand-alone' fruits.

RASPBERRIES

A woodland plant that can tolerate shade, the raspberry fruits primarily in summer and is happiest in a rich soil that errs on the acidic side. There are also autumn-fruiting varieties of raspberry, which extend the season nicely and, despite falling temperatures at this time of year, their flavour holds up well. Raspberries are plants that require little or no attention and are worth growing, even though they take up quite a lot of space. You need at least 1m (3ft) between rows and at least six canes to produce a half-decent yield because once the plant has flowered and fruited it will not do so again that season. As soon as you get them growing well, raspberries will fruit year after year for about ten years in the same place without any problems, given a proper winter mulch of well-rotted manure or compost. It is the successful production of a crop of these delicious summer berries that reminds us of what all the work has been for.

The Victorians grew raspberries in clumps around a central pole to which they were tied. This system takes up less room than the traditional row system of upright posts and horizontal wires that appeared in the first half of the twentieth century, and remains the favoured way of growing raspberries today. There is no difference between pole or row production as far as yield is concerned; it is merely that the latter is what we have become used to. The pole system is perhaps more appropriate for today's back-garden grower, as it takes up less room than the row system. It also works well if you want to grow plants in containers, as long as the container has a depth of at least 30cm (12in) to accommodate the pole.

Raspberries are perennial plants that throw up suckers from their roots, known as 'canes'. The following points will help you understand how the canes grow and produce fruit:

1. Raspberries (except the autumn-fruiting varieties) – and, in fact, all the cane fruit in this section – produce fruit on canes made the year before.

2. Canes appear at soil level at the base of the clump very early in the season and may easily be mistaken for weeds, such as young stinging nettles. Take care. If you hoe off the young canes or pull them out by mistake you will lose the following year's crop of raspberries.

Once you have registered these points, you should have no problems with raspberries. All you have to do is prune the fruited growth in winter and, when this is done, give the plants a good mulch – 10cm (4in) deep – with well-rotted manure or compost. When the fruit start to ripen, you may run into trouble with birds, and as it is complicated to net the plants, some other way of scaring off the birds may be better. Traditionally, soft fruit is grown in a fruit cage to protect it from birds, but this is a considerable investment. If you are lucky, you may get away without caging the fruit.

AUTUMN-FRUITING RASPBERRIES
There is only one difference between the summer-fruiting varieties and those that produce in autumn, but it is a vital one: the autumn fruiters produce fruit on growth made in the same year, whereas the summer-fruiting varieties produce fruit on growth made the year before. At the end of the season, cut back the foliage and canes to ground level and mulch them with well-rotted manure. The following spring new canes will begin to grow, and it is on these that the flowers and fruit will appear in the autumn of the same year.

LOGANBERRIES AND TAYBERRIES
Logans and tays are hybrids between raspberries and blackberries, and while logans are thornless, tays are prickly. The fruits of both are bigger than raspberries and should be left until properly ripe before harvesting. You can tell when they are fully ripe because they will come away very easily from the calyx when picked. The shoots of both fruits can reach at least 4m (12ft) in length

Right: Raspberries are as easy a fruit to grow as exists; with high quality fruit available from both summer- and autumn-fruiting varieties.

in a fertile soil, so they will need some climbing frame to cling to and need tying in to hold the shoots in place. They have exactly the same cultural requirements as raspberries and, like summer-fruiting raspberries, they fruit on shoots produced the previous year.

One of the most satisfactory fruits in this group is the Japanese Wineberry. Very ornamental with spectacular bright red and very sweet fruit, this plant also needs a climbing frame or a set of horizontal wires to cling to. Cut out the fruited wood each year and let the plant ramble up a drainpipe or outdoor stair banister. As long as it gets a good mulch of well-rotted manure each year, it will produce bountiful fruit.

All these berry fruits are hardy in the extreme and absolutely no trouble; they are 'low-visit' plants. It is only when the fruit ripen that you must become more vigilant, or you will lose the crop to birds.

The cultivated blackberry may also be considered as a worthwhile berry fruit. Many varieties are thornless, which makes it stand out from its wild parent the bramble bush, as does the much bigger fruit. The one let down is that the fruits tend to be sour, but you can counteract this by sprinkling with sugar when eaten raw or cooked.

Raspberry Jam

A glut of raspberries should afford you the opportunity to make the essential accompaniment for scones and clotted cream (see my recipe on page 233).

1.8 kg/4 lb raspberries

1.8 kg/4 lb caster sugar

a knob of butter

Makes about 2.3 kg/5 lb

In a large preserving pan, crush the raspberries slightly with a spoon, then heat gently for about 15–20 minutes, until the fruit is soft. Add the sugar, stir until dissolved, then add the butter, and bring to the boil. Let it boil rapidly, stirring frequently, until setting point is reached (about 20 minutes). Test for a set by cooling a little of the mixture on a cold plate (put the plate in the refrigerator first) – if a skin forms it is ready. Skim, pot into warm sterilised jars and cover.

EXOTICS

In the cool temperate zones of the north, perhaps our biggest enemy for growing what might loosely be termed 'exotic fruits' is temperature. Poor light levels also have an impact, but it is largely the prevalence for frost that prevents us from growing so many of the unusual fruits common to warmer climates.

With the help of polytunnels and glasshouses, our scope may be widened, but very often the fruit yields are not perhaps what they might be. What becomes important then is to choose plants that not only produce fruit but have other benefits to make the plant worth growing, such as offering shade or shelter. A passion fruit (*Passiflora edulis*), for example, will produce ample fruit if grown in a sunny room indoors. Not only is the fruit edible but it is preceded by a stunning flower, which doubles the enjoyment of growing this plant. A kiwi fruit (*Actinidia chinensis*), on the other hand, requires not only a high and wide fence for support but at least one plant of each gender for successful cross-pollination. It is also a takes quite some effort to prune and maintain.

OUTSIDE OR INSIDE

Not all of these so-named 'exotics' need to be permanently undercover. Citrus plants, whose natural home is the Mediterranean region, are perfectly happy outside for some of the year as long as they are not exposed to long spells of sub-zero temperatures. For best results, bring them indoors from autumn until spring. They like it cold and dry, not cold and wet, and so, if you wish to grow them, do not overwater them. Their great asset is not only that they flower and fruit at the same time during the winter but also that their blossom is heavily scented. Their fruit yield, however, is minimal unless they get the really hot summers of their native environment.

Much the same is true of melons. An annual plant that is a member of the Cucurbit family, a melon can be grown in a greenhouse or polytunnel and thrive as though it were in its native Iran, but outside it is a different story. Fluctuating temperatures, a surfeit of rain and the never-ending slug problem of the temperate zone, mean that the plant will almost certainly perform below par. Dessert grapes, too, take a lot of effort for what is often a poor yield, unless

'The skill with blueberries is not to allow them to overproduce when the plants are too young, or they will wear themselves out.'

they are grown in a heated glasshouse. If I wanted something that fruits to cover an arbour, then I would choose one of the hybrid berries discussed in Soft Fruit (see pages 179–80), and probably a thornless loganberry.

Two less usual plants that will crop well and are easy to cultivate are figs and blueberries. Ideally, figs like a warm south-facing wall – where I would grow peaches – but they also are very productive when grown in containers in a south-facing position. In pots, their roots are constrained, which they like, and they can be easily brought inside for winter. Blueberries are simple berries to grow and are fast becoming a 'must-have' superfood, as they are high in vitamin C and in anti-oxidant compounds, offering protection against cancers and high cholesterol. Blueberries also grow well in containers. They need very acid soil and this can be created and maintained in the enclosed space of the container much better than it can in garden soil.

FIGS

A half-barrel with drainage holes drilled in the bottom makes the ideal container for a fig as long as it can be moved inside a cool glasshouse or a poly-tunnel over the winter. Choose the less vigorous varieties of fig for growing in containers. During the summer months, figs must have access to full sun at all times to give the fruit a chance to ripen. They like a free-draining compost – homemade compost is excellent – and any roots that protrude from the drainage holes underneath the container must be removed as fig roots are very vigorous. Trees restricted in this way will fruit more readily.

By far the most important point to grasp is that the figs are produced on growth made the previous year. The fruits are made during the previous autumn and will overwinter on the plant.

It takes three years of pruning to establish the shape of a fig tree – goblet or wine-glass shape is ideal. The centre of the tree must be kept 'open' to let in light and air and to prevent disease. Remove the top growth and any dead or diseased wood. In spring and autumn, feed the tree with fresh compost and keep well-watered through the hot summer months. A liquid seaweed feed every fortnight up until autumn will be of significant help in forming the fruit.

BLUEBERRIES

The main requirements for blueberries are acid soil and plenty of moisture, for these mimic the conditions they get in the wild. Also, they need a period of winter dormancy to fruit, so on no account feel sorry for your blueberry by bringing it inside in winter. Blueberries need the cold and must stay outside throughout the year.

Blueberries must establish themselves properly before they should be allowed to fruit. When you buy plants from a nursery or garden centre, try to establish their age. The chances are, you will be sold a three-year-old plant with a mass of fruiting buds. Remove these buds so the plant can concentrate on settling into to its new home (whether a container or open ground). The following year, it will begin to throw up suckers. It is on these suckers that the fruit will appear in considerable quantities. The skill with blueberries is not to allow them to overproduce when the plants are too young, or they will wear themselves out.

Unless you live in an area with distinctly acid soil, with a pH of below six, you will have to grow blueberries in containers. They enjoy this, which is good because if you have any other soil than acid it is impossible to increase the levels of acid sufficiently to fool the blueberry. Ericaceous compost is widely available to buy at garden centres. Plant 50:50 with homemade compost or leaf mould, which usually leans towards the acidic side.

Grow blueberries in full sun, but do not let the compost dry out and make sure that the root zone stays moist. You will be pleasantly surprised at how much fruit they yield over the years, but you must feed the plants through spring, summer and autumn with a liquid feed and 'top dress' the compost twice a year. When it comes to harvest, be sure that they are properly ripe, otherwise you will not enjoy their full flavour. Like the hybrid berries, blueberies should be soft, fully blue and come away easily when picked.

NUTS

Choosing nut trees over fruit trees and bushes is a brave decision in terms of cropping, but there are some beautiful nut trees that do crop well. I have included walnut and hazel, which I consider are the best.

Probably the finest of nuts suitable for northern climates is the walnut (*Juglans regia*). This tree is not available on dwarfing rootstocks and grows to 9–12m (30–40ft). The almond tree, which is closely related to the peach, could also be considered, but it prefers the hotter weather of the Mediterranean.

By far the best nut tree to cultivate is the hazel. This is a multi-purpose tree that performs several tasks for the gardener, the land and the nut fancier. Hazel yields sticks for supporting peas and other plants, bean rods and fencing material and, as such, plays an important part in building the structure of the garden. Quite apart from its exceptional ornamental qualities (catkins in winter are one of the joys of this dour time of year), it is the way hazel trees are sustainably managed and harvested that marks them out as a perennial with special qualities. Traditionally, each tree is cut down only every seven years to produce a quantity of straight hazel rods for use in the garden and on the land. For this reason, hazel can play an important role in a semi-woodland situation or even as a 'low-visit' perennial at the bottom of the garden. It is also a tree beloved of garden birds and other small mammals. Only the squirrel is the hazel's real enemy. They cause damage to the tree and eat the nuts.

WALNUTS

The many assets of the walnut tree include the fact that they are reliable croppers and the fruit are extremely palatable. They are not troubled by squirrels because the fruit have a green toxic outer casing covering the brown shell, and if they fail to produce nuts they offer some of the finest hardwood timber of all, with a distinctive grain and rich colour, for making just about anything.

In my opinion, the fruit is best eaten fresh, whether 'wet' (the shell is still green and has a white nut inside) or fully ripe and hardened off. If you get bumper crops, which is very often the case, it would be well worth considering making walnut oil. The French do. Their walnut oil industry was traditionally based around small farmers with only a handful of trees. Personally, I find the oil too strong and intense and would rather eat the nuts fresh. You will have a walnut tree for life and beyond, so plant it well, in full sunshine, and allow it plenty of room, at least 6m (18ft) from the nearest tree. They make handsome parkland trees with pretty, narrow oval leaves and are much to be admired if you have the space. There are named cultivars on offer, and it will pay to consult a specialist nursery in the region to discuss what might be best. Any normal garden soil is acceptable and they have no specialist requirements except the young tree stems need to be guarded against rabbits.

HAZEL CULTIVATION

Entirely unfussy about soil type, the hazel (*Corylus avellana*) has few specialist requirements. It can be grown in full sun, as an ornamental garden plant, or in shade, where it is equally happy as a low-storey woodland plant just as in its natural environment.

Give the young hazel trees plenty of room, at least 2m (6ft) between each, but bear in mind that it is more than possible to only 'part coppice' a tree, so you can manage with only one plant. The nut yield, unfortunately, is likely to be minimal, but hazel stakes, canes and supports are expensive to buy and, if you can grow your own, it is satisfying on many counts. You could grow bamboo and have an edible harvest of the shoots as well as canes (this is something that may well pick up as a commercial crop in Europe before too long), but hazel poles are stouter and the plant more desirable generally.

HAZEL COPPICING

Traditionally, hazel was coppiced over a seven-year cycle, and still is by those with more than seven plants. However, this is not strictly necessary and, as I have mentioned above, a single plant may be 'part coppiced' every few years without compromising the production rate of the tree. The key is to let the plant establish itself first for at least four to five years before beginning the coppicing process. Once hazel rods of a suitable size have grown into the shape and size required, a percentage – say 30 per cent – may be cut down. This will leave enough on the plant for it to continue growing. New growth will appear from the 'stool' the following season and the cycle continues with older growth cut out each year and newer growth allowed to grow on for three or four years. On a seven-year cycle, the entire plant is cut to a stool roughly 20-30cm (8-12in) above the ground and allowed to regenerate for seven years.

WILD FOOD

FORAGING

I don't know many people who are completely self sufficient in their food, but we all know someone who forages for wild food even if it is something as simple as collecting sweet chestnuts to roast over the fire.

For me, this is where the value is – supplementing our diet with food that is not available to grow or buy. More than this, gathering wild food requires a total interaction with nature to procure it; walking a hedgerow to pick some wild herbs or gathering some edible seaweed or some mussels from the shore. It sounds dreamlike, but the reality is that if you can be bothered there is plenty out there and it's all free.

Gathering from field and hedgerow appeals to me greatly because it presents an opportunity to understand how your local flora, fauna and the seasons work. Gardening, as I have mentioned before, is beset with regulations and conditions. Foraging is not; you just have to be there at the right time.

For those in urban areas, there is clearly travel involved, but once that is overcome, the honing of the senses to be on the lookout for edible morsels leads to a better understanding of soil, plants and how they grow and interact together. Observation in both the garden and the country has a symbiotic effect; the two are mutually beneficial. This, I feel, is important and can start many a child on the path towards a love of nature and, ultimately, gardening. For those of us who already garden, we would do well to learn as much from nature as we can.

WEEDS AND OTHER GREEN PLANTS

The search for wild green plants for cooking or eating raw must start for most of us in the garden. Whatever we are growing in the garden, we cannot keep the 'wild' out, and it is here that we can begin to make that vital connection with the wider and wilder side of nature. Once we begin to realise that there is more to a weed than it simply being a 'plant in the wrong place', we take our first steps on the journey towards the meeting of gardening with nature.

NETTLES (*Urtica dioica*)

Top of the 'most-wanted list' is without question the stinging nettle. A prolific perennial weed, the nettle should be welcomed in the garden not only for its edible qualities and as a host for various insects but its presence indicates that the soil is rich in nitrogen. It is what is known as a 'dynamic accumulator', a

Right: Nettles are both edible and a good indicator plant for determining the health of the soil, both in the garden and the wild.

Recipe for Nettle Soup

Wearing a thick pair of gloves, pick only the tips and young leaves of the nettles. Be absolutely sure it is a virgin patch of nettles growing away from where animals, such as dogs, cats and particularly rats might find them. Sweat a chopped onion in a good dollop of butter and then add the nettle leaves whole. Put the lid on the saucepan and carry on sweating the nettles until they are cooked. Add a ladle of chicken stock and carry on simmering for a few minutes. Then add further stock to make the desired quantity of soup. Simmer for twenty minutes. Whizz up the liquid and nettles in an electric blender and serve with a swirl of cream and a grate of nutmeg..

plant that gathers nutrients and indicates the presence of those nutrients in the soil. It follows, then, that these 'high-nutrient foods' must be good for us. Nettle soup has long been famous as a 'famine food', sustaining countless families in times of hardship, conflict and so on.

GROUND ELDER (*Aegopodium podagraria*)

This ubiquitous perennial weed is tough to get rid of in the garden even though it has a fairly shallow root system (unlike nettles, docks and dandelions, which put down tap roots that are hard to dislodge). From an edible perspective, the advantage of ground elder spreading fast is that it offers a good, regular supply of fresh leaves to harvest.

You may be wondering why on earth anyone would want to pick ground elder when they probably have a plentiful supply of lettuces, rocket, dill and other succulent salad ingredients to choose from in their vegetable garden. The answer is this: it is rewarding to feel you are coming to know the plants and weeds that grow both in your garden and the wild and to know what they are, why they grow, what is in them and whether they are good for you or not. Surely this is the bedrock of eating anything – does it taste good and is it good for me? I am not for one moment saying that you should try eating a salad made of pure ground elder, but what I am suggesting is that a leaf or two of this common weed may open up a whole new range of possibilities and a new way of thinking for you. Ground elder is fairly decent raw, but the leaves do taste better cooked and perhaps added to the nettle soup recipe (described above).

CHICKWEED (*Stellaria media*)

Chickweed is a common biennial weed which is very tasty raw. It tends towards the stringy side as it gets older and is best eaten young, so weed out the plants in early spring before it flowers and sets seed. Similarly it can be eaten in autumn just after it has germinated. Chickweed has a gloriously earthy taste and looks decorative in the salad bowl. Break up the lengths of stems between the small leaves, as they can taste sharp. The flavour resembles that of spinach with a bit more of a hint of iron.

DANDELION (*Taraxacum officinale*)

Deep rooting and tough to shift, it is best to look upon the dandelion with benevolence and as an addition to the garden rather than as a hindrance. Once cultivated as a salad leaf, it is very easy to do the same in the garden today with dandelion, as was done by gardeners a hundred years ago and more. Choose a vigorous dandelion plant and twist or cut off the green leaves at the base of the plant (these are far too bitter for our tastes anyway). Cover the crown of the plant with a flower pot and put a slate or stone over the bottom of the upturned pot to stop any light getting in through the drainage hole. In two weeks or thereabouts succulent, translucent white leaves will start appearing from the crown. Put the plant through two or three croppings, as you would the perennial vegetable sea kale, and then dig the plant up and compost it. You will find the root much easier to remove because of the tough regime it has been put through. Cultivated dandelions are becoming popular again amongst salad growers for annual use. A red-ribbed variety has had much of the bitterness bred out of it.

FATHEN (*Chenopodium album*)

A very common weed with unmistakably pointed leaves like tiny Cornish shovels, this is delicious raw when served with other mixed salad leaves.

HEDGEROW

Next time you pass a field of dairy cattle, stop to consider why so many of the beasts brave the electric fence and stick their heads through it to graze the hedgerow. This is because the glossy green fare to be had in the field is nutrient poor, chemically fed grass with little or no goodness in it. Rather than taking nutrients from the soil, it is being fed artificially with nitrogen for the purposes of quick growth. The ruminant knows what it needs to nourish itself and so goes looking wherever it can to find the nutritional and medicinal herbs it requires. It finds them in the diverse flora of the hedgerow.

The common weeds of the garden are the route into the hedgerows; once you have come to realise the potential of the wild garden plants we know as weeds, and that many of them have good flavour, you will be keen to expand your wild collecting. Any walk in the country suddenly becomes a reason to make wild garlic soup or ponder whether a certain mushroom is edible or not. Like as not, you will come across much in the way of edible food in the country beyond the garden, but be sensitive in removing it from source. Use your judgement carefully as to whether you can justify picking it and at what cost. Below, I have listed one or two of the better tasting wild plants that are usually found in abundance and can stand picking, discovering them may help to widen your scope.

HOPS Many wild plants that are eaten as greens or raw as salad are unexceptional, but there are some which are highly sought-after. The wild hop (*Humulus lupulus*) is very classy. There comes a point in early summer when the snakelike shoots of the wild hop can be seen clambering all over hedgerows. Dark green and looking for all the world like bindweed before the leaves have unfurled, they are strong in flavour, and rich with the taste of chlorophyll, almost to the point of bitterness. Lightly steamed and added to a salad they are a supreme wild shoot.

RAMSONS (*Allium ursinum*) Another springtime delight, the wild garlic smell of ramsons is made all the more alluring by the abundance of the plant. When you come across wild garlic, it is often a huge aromatic carpet of white flowers and fresh green leaves taking over big slices of a wood. The leaves, which are the edible part, tend to lose the garlic flavour in cooking, but the texture is velvety and heavenly like young annual spinach.

'It is rewarding to feel you are coming to know the plants
and weeds that grow both in your garden and the wild.'

SEA BEET (*Beta vulgaris.* subsp. *maritima*). If you are fortunate enough to live close to the sea, this wild ancestor of spinach beet is an outstanding wild plant. It has a taste of the wild that is simply not there amongst the newcomers that have been bred from it. It is not as sweet as spinach beet (perpetual spinach) or swiss chard, but it is far richer and makes excellent soup. It can be found on cliffs and in lanes and hedgerows in coastal areas, where it grows freely as a biennial, so if you are picking leaves avoid those from plants that are going to seed in their second and final year. Two other maritime favourites are marsh samphire and sea kale. Samphire takes some finding on salt marshes while sea kale will not be found in edible form on the shore although you can gather seeds and propagate new plants for forcing (see Perennial Vegetables).

FRUITS AND NUTS

Some of these are perhaps our most likely targets when we think of gathering wild food. Sweet chestnuts (*Castanea sativa*) for roasting; sloes (*Prunus spinosa*) and their larger and sweeter relations, bullaces (*Prunus domestica*), for adding to gin and vodka; or even picking away at the often empty shells of the beech nut, beech mast (*Fagus sylvatica*). Elderberries (*Sambucus nigra*) make wonderful drinks and exceptional fritters while there are some lesser lights that can be put to excellent use.

Rowan berries (*Sorbus aucuparia*) were a famous famine food of the Scandinavians in hard times, as they are very high in vitamin C and store well. They make good chutney. Rosehips (*Rosa canina*) are similarly high in vitamin C and have a handsome sweetness to the outer flesh. The Guelder rose (*Viburnum opulus*) may be a plant that will startle you in the autumn with its vast quantities of scarlet fruit that resemble larger redcurrants. Easily identified from the maple by shape of the leaf, the fruit needs to be cooked to destroy the toxins so juice, jam or jelly are the best bets.

The bramble (*Rubus fruticosus*) has been discussed in Cane Fruit (see page 180) in the shape of the cultivated blackberry, but the crab apple (*Malus sylvestris*) is certainly worth a mention, for it makes superb jelly, often with an unmistakable pink flush. Seek this tree out if you can, for the wild crab makes jelly like none of the named cultivars, even though they do excellent work in pollinating domestic varieties of apple, and use your powers of observation to do so. Be still, listen to the sounds of nature around you and admire its wildness from speck of soil to grand old crab apple.

Right: Sloes are one of the finest hedgerow fruits for use in wines and spirits.

Overleaf, page 200: Seaweed is one of the best freely-available fertilisers anywhere, if you are lucky enough to have access to it. The only requirements for collecting seaweed are a fork and a means of transport. The good news is that the salt is not harmful to crops, apart from the tiniest of seedlings.

FUNGI

Far less hazardous than we are led to believe, collecting fungi is probably best taken a step at a time. Identification can be difficult when the mushrooms are no longer young and fresh and are starting to break down. Obviously, if in doubt, pass the fungus over. The field mushroom (*Agaricus campestris*) is widespread in pasture, as is my top choice the giant puffball (*Lycoperdon giganteum*). Catch them young at around the size of a cricket ball, and they are pure cream: thick and intense and tailormade, like all mushrooms, for bacon at breakfast. In the woods, the cep (*Boletus edulis*) reigns supreme. This is a rich mushroom with a live almost smoky flavour.

SEASHORE

Picking mussels, going fishing, shrimping, or digging for cockles is all part of life by the sea where there is a wealth of food that can be gathered. Also, no trip to the sea is complete without a seaweed harvest. Dulses, lavers and, in particular, sea lettuce are eaten all over the world, constituting a major part of the diet in some countries.

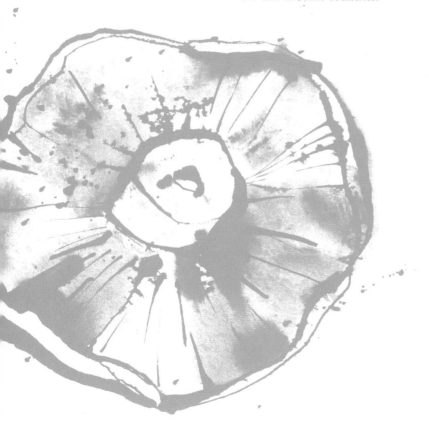

Shrimping

Coastlines all around the world hold their own secrets. Inshore waters are havens for fish, crustaceans and edible creatures of all shapes and sizes. Children have always been fascinated by rock pools and beaches, and I still am.

Wild food from the seashore is often abundant and takes only some effort and a little time to be able to identify it with confidence. Like wild plants, most of the edible morsels are never too far from the surface and, for example, freshwater mussels, whelks and even oysters can be found on riverbeds at low tide.

In the UK and elsewhere, crabs and lobsters are most commonly caught in pots that are baited and lowered by rope to the seabed. There are other methods of catching these shellfish, particularly crabs, when their favourite hideaways are exposed at low tides. Lobsters don't present such as easy challenge, as they rarely leave the safety of the weed cover on the sea floor. Edible crabs, on the other hand, the ones with the brown bodies and black claw tips, like a deep crevice in a rock, and these can sometimes be uncovered at low tide. Overhanging rocks are a good place to begin looking. Get down on your knees (a wet suit is always good for protection, as are rubber shoes or trainers) and you will see, lurking in the depths of the crevice, two shiny eyes. All you need then is a strong but thin iron bar with a gaffe or sturdy hook on the end to extract the crab.

Shrimping or prawning at low tide when plenty of rock pools are exposed is an altogether more leisurely pastime. A sturdy net and a strong canvas shoulder bag in which to carry the quarry is all you need. Don't use a bucket, as you will only have to put it down somewhere. When the tide changes and begins to come in or 'flood', the prawns come in with it. As they do so, they move up out of the sand into the open. Once exposed, the prawns can be picked off one at a time in an activity that in Cornwall we have always known as 'stalking'.

However, patience is the watchword. Rushing or excitement will not bring rewards. Any sudden movement will send the prawns tearing backwards and downwards at breakneck speed and you won't see the next one for several minutes. The only way to catch prawns is to slowly ease your net under the approaching prey. Then, when he is standing fully on your net, lift it, but very gently. It is a very slow manoeuvre.

Sea beet soup

There is something highly satisfying about a green leaf vegetable soup. Nettle is one of my favourites. It just has to be good for you, brimming as it is with vitamins and nutrients. Freshly gathered and cooked is the best way to get maximum flavour and goodness from leaf vegetables because the moment they are cut from their parent plant they begin to lose their nutritional value on account of cellular breakdown.

Sea beet, which is the parent of spinach beet (perpetual spinach), Swiss chard, beetroot and sugar beet, has a flavour that outstrips all its junior relations. In the UK, sea beet grows wild in coastal regions and its low-growing rosette shape is easy to identify. It is a biennial, so if you find it in the spring of its second and final year, it will have long stalks with flowers or seeds. Although the wild plant does not grow to the size of cultivated spinach beet, there are always plenty of leaves to pick without damaging the plant. It is not really necessary to wash the green leaves you have gathered unless they are very muddy. A little bit of grit never hurt anyone.

Sea beet is, for my money, much better in soup than as a stand-alone green leaf vegetable. To make sea beet soup, the method is the same as for spinach soup. In a large saucepan, place a chopped onion and sweat it in butter with a celery leaf or two. Put two large handfuls of sea beet leaves into the saucepan. Add a chopped potato and sweat that too. The mix should then be sweated further until the leaves have collapsed. Add a ladle of good vegetable stock and cook gently for 10–12 minutes before adding more stock to make up the of soup. Simmer gently for a further 30 minutes. Whizz up the soup mix in a blender and reheat if necessary. To serve, add a swirl of cream or crème frâiche to each bowl and a light grating of nutmeg. Alternatively, this soup can be made with any other green leaf vegetable.

ANIMALS

INTRODUCTION: ANIMALS

While the intensive cultivation of plants can be seen as an efficient way of producing fresh food, the subject of intensive rearing of animals for the same is more controversial. Our access to well-produced animal products has widened and increased, but we must also realise that the presence of animals as a means of soil fertilisation is important, if not essential, to both our own and our land's health. Of course, millions of people the world over live well on a meat-free diet and animals do not graze their land, but it is not being pedantic to point out that cycles of life in nature, and particularly soil, are entirely dependent on animals, even at a base level in the form of insects, worms and other small creatures. The homemade compost that feeds your plants is manufactured to a large extent by animals. Even if you do not use compost on the land, animals will play a significant part in the fertilisation of it. Whether or not you choose to include animals in your diet is a choice. My point is that in whatever form they appear in nature, animals are beneficial and help to create a balance. It is only man who has upset this balance by the excessively intensive farming of animals. Just one of the harmful by-products of animal rearing is huge quantities of methane – a damaging greenhouse gas.

The importance of diet and nutrition has been discussed in Chapter One, so now let us consider which animals it is possible to successfully rear in a small area. We will also look at the issues involved when raising animals, how animals affect us, and what they contribute to our lives and our land.

SPACE AND TIME

How to best use one's time and make the best use of available space are issues that concern most smallholders. Home-reared animals and birds need looking after and they need some land, as do the plants growing in a productive vegetable garden. The design and layout of your site becomes particularly important when you introduce livestock into the mix, as does how the site can be arranged for the good of all.

Ethics and welfare, two words that are applied more and more regularly to animal husbandry today, must be a given in the treatment of animals in our charge. If we choose to rear animals ourselves, we can and must raise them humanely. It is important for our own sense of well-being to respect their lives and to know that they have been well-treated before they are slaughtered. No matter how rational you are, the despatching of a well-loved ewe or a calf that has fed from your hand can be very upsetting. Emotions surround the issue of raising and slaughtering animals, so you need to know what you are getting yourself into from the start. Can we then justify eating our animal friends, or is this too simplistic a view? Animals do bring companionship and whilst the aged rooster who escapes from the pen and scratches up several lines of seedlings in the garden may be the subject of wrath and abuse, he is still the rooster that wakes us up every morning, and though probably only good for the pot, he more than likely has a name. This is not sentimentality, it is life.

THE BENEFITS

Each group of animals and birds reared by the smallholder can contribute food and other beneficial by-products. (These are discussed in detail, under each separate animal.) It is my experience that animals around a smallholding of any size, are an essential part of the life cycle of not only the land but also the human consciousness. No one says that you have to eat these animals, but the intangible benefits of just having them around are legion. Nowhere is this more important than in soil fertilisation, where the animals' relationship with the land is beneficial to both parties.

If the intention is to grow vegetable crops alongside raising animals, the potential to produce food for the household is great, while the vegetable and animal waste in the form of compost and manure will feed the land; the two go hand in hand. Obviously, the smaller the space, the harder it becomes to rear larger animals, but the very presence of an animal on a site, however great or small, will mean the production of waste material in the form of manure. This is advantageous for a number of reasons: firstly, we can use the manure to feed our own soil; and secondly, we do not have to buy in or waste energy or fuel transporting manure from elsewhere.

The debate surrounding the provenance and traceability of food has come to the fore in the last decade due to increases in food and hygiene scares, especially concerning animals. The devastating outbreak of Foot and Mouth disease at the turn of the millennium and the appearance of mad cow disease and its human form variant CJD have led to concern over the safety of meat products from cattle. However, somewhere near your home there will be a farm that produces high quality products to stringent standards. Nevertheless to be absolutely certain you know what you are eating, you have to grow whatever it is, animal or vegetable, on your own ground. Only by feeding your animals on land and food that you have grown yourself is it possible to do this.

POULTRY

Nothing conjures up the image of the rural idyll more than a few hens scratching around the yard. But if you don't have a yard, it is possible to manage poultry effectively in a small space because these birds are tough, self-reliant and wise. Poultry are highly enjoyable to keep, and though they can bully one another, they resort to such behaviour only when stressed and unhappy. Given good husbandry, they are content and will fulfil their potential.

The prime requirements of poultry are a varied diet, ground that they can scratch up for grit (to help shell production and digestion) and use to enjoy a dust bath, a plentiful supply of water, shelter from wind and rain, shade, protection from predators both day and night and somewhere to lay eggs. Sounds a lot in the quest for the breakfast egg, but there is more to it than that: for the 'sufficient' practitioner, the humble hen is great news.

REASONS FOR KEEPING POULTRY

It is important to take into account all the possible reasons for keeping poultry because, easy as they are, they need to be looked after. They cannot be abandoned. If we are thinking along the lines of keeping hens only for their eggs, we are missing a trick. These birds are multi-taskers and, if we give them the right conditions in which to work, they will produce much more and are a very valuable commodity.

The direct yields from poultry are eggs, meat, feathers, manure, more poultry and human companionship. The less obvious which are of great benefit to the gardener include pest and weed control, ground clearance and soil fertilisation, and a use for kitchen scraps, which are then converted into rich poultry manure by the chickens. Of course, most kitchen scraps can be put directly on the compost heap, but if they are given to poultry instead, they will feed the bird and help make eggs, meat, feathers and so on, and manure.

Although a less obvious yield, energy from each bird's body warmth is an additional by-product of keeping chickens. During the cold nights of the winter months, it is worth trying to arrange a system whereby the birds are moved into your greenhouse or polytunnel when the temperature drops overnight. Their bodily warmth will be released slowly into the space and help you save on heating bills. This release of heat into your greenhouse or polytunnel will help you propagate young seedlings that need warmth.

Right: The humble hen has a huge amount to offer in terms of yield.

HOUSING

If at all possible, all animals kept on a smallholding should live outside. Poultry, on the other hand, need protection at night and enjoy coming into the hen house at dusk and stay inside for the dark hours. I once came back home in the late afternoon to find our hens sitting on the sofa, watching TV.

In whatever structure you house poultry, it needs the following: roosting perches and nesting boxes with roofs that are easily accessed for egg collection. The house must be solid to prevent attack from badgers, which I have known to get into hen houses at night and cause mayhem. (Foxes are more of a daylight problem and don't have the strength to break into most hen houses.) Rats are also extremely common. They are sure to appear where food is in evidence, Rats are egg thieves by nature and they like the cover provided by a hen house and will soon set up home beneath it. This will be unwelcome, particularly in an urban environment, so at the first signs of rats, act fast and set a humane trap. Poisoning is not an option, as the knock-on effects in the food chain are untraceable.

FEEDING

Organic and free-range eggs are what most of us look for when buying eggs, and this is precisely what we can achieve by rearing our own poultry. In the process of providing fresh eggs, we can have the birds work for us at the same time. A significant part of a hen's diet can be gleaned from the soil, preferably where there is grass or plant matter available. The time-honoured tradition is to keep chickens out of the productive garden, but with the exception of freshly sown seedbeds and rows of young seedlings, they do not cause much damage, especially if they are moved around the whole site regularly. This is the real definition of free range, and as long as their diet is supplemented with grain, they will not cause wholesale damage to the cultivated areas of the garden.

If free-range birds are too much of a risk, and in small spaces it may well be, the best solution is a form of moveable pen or electric fence. If a number of birds are penned into an area for a length of time, they will derive nutrition from and fertilise the ground for the greater good of themselves and the garden. When chickens are moved about the land in this way, they must have access to water and shelter from wind and rain, even if the latter comprises a small covered area under a sheet of galvanised metal. Chickens are more self-reliant than we give them credit for, but they must be kept safe from predators.

After some days, the chickens may be moved on to fresh ground, where a new round of weeding and pest control will take place and all that will be left behind is some wonderfully nitrogen-rich soil ready for cultivation. Hens are excellent for dealing with perennial weeds as long as they are scythed down before putting the birds on the ground. Brambles and nettles will also be much easier to dig out after the birds have scratched around the root zone and pecked away at the stems.

BREEDS

All of our domestic breeds are derived from the Jungle Fowl, a magnificent bird that roams the forests of Asia. The choice is wide, but local knowledge is everything, so take advice from the people in your area who raise poultry. Of the more popular breeds, Marans lay beautiful brown eggs, Light Sussex are good dual-purpose birds and Rhode Island Reds are very reliable. Black Rocks are another suitable breed for their superior egg size and dependability.

PREDATORS

The fox is the prime offender for his brazen daylight attacks and excess of cunning. Running a rooster with the flock will not only provide fertilisation for the eggs, but will also give some protection from the fox by way of an early warning signal. Roosters in the town may not be that popular with the neighbours, but they will act as a deterrent. Foxes are unwelcome in built-up areas, but they seem to like it there, as it suits their scavenging mentality. The fact that they often look in poor health gives them more incentive to attack chickens, so you have to be very vigilant. As we all know, foxes are clever opportunists and will wait to strike when the going is good. There is little we can do to stop this. However, you will know when there are foxes close by because of the strong scent that they leave. The presence of dogs is a good way to keep foxes out, as is a robust gander, who will happily co-exist with the poultry. The fox may be fond of goose, but this fellow has the best early warning system of all and will put up a spirited resistance.

GEESE

Hopefully, by now, you will have lost the desire to have a perfect lawn somewhere on your patch. If you intend to keep geese, this is just as well, as they will head straight for your grass. Geese are good grazers and will keep grass down, but the verdant strip will soon become littered with guano. Penning is not the answer. Whilst it affords protection, these large birds need to be free range and they need access to grass.

You need to give some thought before embarking on the road to goose rearing. These are large birds, often with an aggressive, territorial attitude. A flap of the wings or a lunge with a strong beak can hurt a child, and, by nature, they can be bad-tempered. They also live for a considerable length of time, sometimes as much as 20 years. The upside is that geese are a deterrent to foxes. They marshall poultry well and can also act as a good intruder alarm with their loud honking. They must be housed at night, but they are very hardy as regards climate and, unlike ducks, they do not need water for swimming, only a clean supply of drinking water.

If you intend to breed from your geese, you will need a pen to protect the young goslings. The goslings will need to be fed on milk and bread until the time when they learn to eat grass. Like all fowl, the more space, the better. If you are fortunate enough to have an area put down to trees with long grass, let us call it an orchard for the sake of argument, then this is the place for geese.

ROTATION

Most fruit trees and ornamental trees will thrive in grass as long as it is kept at least 30cm (12in) in diameter from the trunk when the trees are still young to prevent the grass from competing for nutrients with the young tree roots. To manage the manure produced by the geese and to allow the grass time to regenerate between grazings, the best thing is to section off areas of the orchard and allow the birds in only one area at a time. Leave the birds to graze for a few weeks or however long it takes them to crop the grass very short. After that, move them to the next area, and you will find that the grass will grow back quickly. If the area has been badly flattened by the geese's webbed feet and the grass looks very unhappy, then it is a good idea to let in poultry to scratch up the ground and cancel out the compaction caused by the geese. Alternatively, you could rake over the surface of the soil to allow air in for the grass seed to germinate, or sow with new grass seed.

THE GOURMET ASPECT

Despite the fact that geese are rather fierce and often quite scary, to be perfectly honest, the good outweighs the bad for me. Geese pull their weight. They help the other fowl and they will keep intruders, especially of the human kind, away. You must keep on top of the rats, who are very keen on goose eggs, but if you do manage to raise a clutch of goslings, they are worth their weight in gold. You don't often see geese on the market other than at Christmas, so the goslings will pay their way handsomely if brought to maturity. Once you have killed a goose for the table – and you may well not be sorry to see it go – the meat and the by-products are of a very high quality. As too are the soft pure white down feathers. There is nothing quite like the comfort of having a pot of goose fat in the fridge, especially when you have a fresh egg to fry in it.

DUCKS

When it comes to keeping fowl, my choice would be poultry and geese together, not ducks, as they need a good stretch of fresh, running water. Unlike geese, ducks really do need water for swimming, just as a hen needs somewhere to have a dust bath. Water is essential on any smallholding, but a duck pond needs running water moving through it, otherwise it will become stale and stagnant, making it unhygienic and foul-smelling. Rainwater falling on the pond or filling up the pond from time to time from your water supply will not be enough to prevent the pond water from stagnating.

The exception to this is when you have only one or two ducks. A pair of Khaki Campbells will hammer the slug and snail population in your garden and fertilise the ground as they go. They will lay eggs happily given a safe house in which to do so and they make a great sight strutting around the place with their long necks and heads held high. But any more than a couple of ducks, and before you know it the place has become paddled down by the ducks' web feet and ruined.

Khaki Campbells and Indian Runner types are excellent for gardens and small spaces. They pose little or no threat to crops and are good egg layers. Puddle ducks, as they are affectionately known, such as the Aylesbury or the Muscovy, are reared more for their meat. They are heavier and not particularly mobile. The benefit of these types is that they produce excellent feathers and down as well as meat.

PIGEONS

Pigeon meat and eggs played an important part in the human diet centuries ago, as did peacocks and swans, at least amongst the nobility. Today, the appealing side of pigeon-rearing for meat and eggs (and feathers) is that they are fairly low-maintenance, as they are the only birds that spend a good deal of time in the air rather than on the ground. This reduces the threat of predators, and pigeons really only fall prey to highly mobile falcons, such as the sparrowhawk. Pigeon eggs and chicks are taken by squirrels who are more carnivorous than some give them credit for, but it is possible to keep the threat from rats and cats away by placing your dovecote on a high pole or wall-mounted on the side of a building. The birds may be fed corn and bread on the ground and the only regular visit that need be made via a ladder is to clean and freshen up the water.

GETTING STARTED

There is very little equipment needed. All you have to do is buy or build a dovecote to your specification and purchase some birds of your choice. It is possibly best to avoid pure-breds because the nature of keeping pigeons is that they will attract wild pigeons who will cross with your beloved pure-bred fantails and spoil the bloodline. When the birds first arrive, you must keep them in the dovecote for a couple of weeks so they come to know it as home, otherwise they will leave and not return. Tack chicken wire to the outside of the arches in the dovecote to stop them from escaping; they will not mind being house-bound for a short while. Give them perches inside to roost on and do clean out the dovecote at least once a week to stop the build-up of pests and diseases. Use straw bedding on the floor to make cleaning easier and also for nesting. The pigeon guano is nitrogen-rich – as good as that of poultry – so remove the soiled straw to the compost heap, where it will act as an excellent activator and hasten the rotting of the heap.

THE PRODUCT

Pigeons eggs are delicious, but the main idea of keeping pigeons on a smallholding is to eat the young birds, known as squabs, when they are about seven to eight weeks old and before they begin to breed. The meaty pigeon breast is the obvious target, but the carcass is also excellent for making stock. Beware, though: the pigeon bones are very small and easy to choke on.

BEES

If the cultivation of a crop without the culling of an animal is more appealing, then beekeeping is for you. It is a hugely rewarding pastime with no end of benefits, including fruit pollination in the garden and orchard, and honey and beeswax. As a behavioural study of living creatures, beekeeping is without compare. The way the hive is organised and how the honey bee goes about its life and – more importantly – the life of the colony, is utterly fascinating.

I have been a 'hands off' beekeeper for years. I started when the apple, pear and some plum trees in our orchard failed to set fruit properly for two years running. I resurrected two ancient hives that I found languishing in the attic. I then imported a nucleus of bees from another beekeeper into one of the hives, and by midsummer they had swarmed, allowing me to introduce the swarm into the second hive. The resulting pollination and fruit set the following year was astonishing and I was hooked.

Fine, calm weather at blossom time undoubtedly plays a significant role in the pollination of fruit trees, but something about the presence of the bees (you would never know the bees were there unless you found the hives) felt right. Open-pollinated plants depend on insect activity as well as wind and rain for fertilisation, and bees have a huge responsibility in this area. I made it my business from Day One to take only a small amount of honey off the hive at the end of the season. After all, how fair is it to rob the bees of all their hard work and feed them a substitute sugar solution? I know now that bees need only a third of the honey they have made to survive an average winter, but nonetheless it seemed only reasonable to harvest a small amount.

My ignorance surrounding bees stretched back to horticultural college. There, the commercial tomato glasshouse functioned with artificial lighting and heating for twelve months of the year, and relied upon imported bumble bees for pollination. In the wild, the bumble bee is not present in sufficient quantities to carry out all the work needed for pollination, so a hive in the garden will increase your chances and reward you with good yields of fruit.

BEEKEEPING ETIQUETTE

The position of the hive is of the utmost importance. It should be in a warm and sheltered location. Bees are like newborn lambs – they dislike cold winds and rain, particularly when they occur together. In spring, if the conditions are

wet and windy, they will stay in the hive and will not pollinate flowers. Place the back of the hive against a wall or fence, where the warmth of the sun can be trapped and bounce off the surface. A protected site like this will continue to distribute its heat into the night when the sun has gone down and will warm up the area first thing in the morning, encouraging the bees to leave the hive and get to work. Make sure the hive is placed on blocks off the ground and put a heavy stone or round of wood on the roof to stop it blowing off in the wind.

Never walk in front of the hive. This will take you across the flight path of the bees and is the most likely way you will get stung, not because the bee sets out to sting you but because she will become attached to your clothing or hair and have no other option. One sting, and the honey bee dies. As a general rule, leave the hive alone. Bees dislike disturbance and whilst it is tempting to open a hive to see what is going on and how the honey making is progressing, try and avoid it; the balance of the hive will be upset. When you do have to carry out a task, do so on a warm, sunny day, at the end of the day, when the bees are less active. It must be warm, as bees do not like the cold.

TOOLS AND EQUIPMENT

First and foremost, you will need an all-in-one bee suit that covers you from head to toe and a sturdy pair of gloves. The bee's sting can penetrate surprising layers. Wellington boots are good as long as the suit fits snugly around the ankles. Next is either a screwdriver or a specialist hive tool for opening, working inside the hive and scraping off excess wax. Finally a smoker. Smoking the bees with wood smoke drives the bees down inside the hives to load up with honey in case they have to flee. This makes working in the hive easier and helps to avoid crushing bees when you add to or replace layers of the hive.

ESSENTIAL TASKS

Assuming you have started from scratch, that you have a hive and the bees have been introduced in the form of a bought nucleus or swarm and have taken to the hive, then you are under way. As summer progresses, the bees will work tirelessly towards making as much honey as possible. Two things: extra layers, known as 'supers', have to be added to the hive in which the bees deposit the honey onto frames. These thin rectangular frames hang down next to one another in the 'super'. Secondly, the queen bee must not be allowed up out of the brood box (where the eggs are laid and will hatch). A 'queen

Left: Bees are essential for any holding, especially where there are fruit crops under cultivation.

excluder', a thin plate of metal with holes must be fitted between the brood box and the 'supers' above. This allows all the worker bees access throughout the hive, but not the queen.

SWARMS

When the hive becomes overcrowded or the 'workers' (all of whom are female and make up the bulk of the colony) decide it is time to install a new queen, the hive may swarm, leaving the old queen and some of her colony behind. This can cause damage to the hive or it might be a peaceful process. What is hoped for is that the new swarm will fetch up somewhere close to the hive, allowing you to capture it and introduce it to an empty hive that you have nearby. This will turn your single hive into an apiary.

A swarm, whilst looking very threatening, is nothing of the kind. The bees, which hang down in a huge living blob, the size of a large watermelon, have only one thing in mind: to stay in a tight group around the queen. To capture a swarm, all you have to do is gently wipe off the bees into a large cardboard box and turn it upside down. Prop up the cardboard box and leave it open a little bit to let the missing bees know where the rest of the family is. The next day, open the box fully by the entrance of the empty hive and watch in amazement as the bees first examine the new premises and then stand on the lip of the hive and buzz their wings to encourage the others to come and take a look – a process called 'fanning'. In a couple of hours, they will have taken up residence and, if you are lucky, you will see the queen being serenaded into the hive. It is a wonderful moment and one that will draw you deeper into the extraordinary world of bees.

GATHERING HONEY

This is a matter of personal choice. I am a huge fan of raw comb honey and, as such, prefer to take the occasional frame out of the hive from late summer through autumn. In my experience, comb honey does not store well, so you are duty-bound to eat it up yourself, or share it and barter it with your friends and neighbours. Others prefer to separate the liquid honey from the wax by putting the frame in an extractor. This extractor spins the frame at high speed and takes the honey out of the comb, enabling the liquid honey to be jarred up. This extraction method first requires the removal of the sealing wax from the outside of the frame, which is best done with a hot knife. This wax has many uses, as does the honey. The frames can then be placed near the hive where the bees will clean them off before they are used again in the hive, although some frames may have to be re-wired first.

VAROA MITE

This is a mite that attaches itself to the underside of the bee and, like an aphid on a plant, quite literally sucks the lifeblood out of the insect. It has put bee populations under severe threat in recent years and is a very serious problem. A varoa mite on a bee is the equivalent of a person having a large edible crab attached to their chest. At present, chemical treatment seems to be the only solution, but it does work. We can help to avoid this mite with good husbandry. Make sure that the hive does not become overcrowded; add more 'supers' to increase the room in the hive; and ensure that dead bees and other debris are cleared out of the bottom of the hive, which should be equipped with a holding tray. The bee is a clean insect by nature and insists on high levels of hygiene in the hive, but this mite is persistent and untreated hives will soon succumb.

BEEKEEPING THE SUFFICIENT WAY

Beekeeping can be enormously time-consuming because it is so absorbing and activities around the hive cannot be rushed; bees are very sensitive and so you have to work calmly and quietly with them so as not to upset the balance of the hive. Call me lazy, but my 'hands-off' attitude towards beekeeping and the harvesting of the product seems to be the right way. It is a 'live and let live' attitude. Let the bees do their thing; they are helping your garden and encouraging its diversity through pollination of flowers and fruit. Talk to them and encourage them as you would any other animal; they respond. In my experience, some bee colonies are friendly and some are not, so you need to make an effort. Finally, reward the bees by leaving as much of the fruits of their labours where it is supposed to be – in their hive, where they can consume it at their leisure and prosper.

GOATS

The bad boys of the animal world, goats are too often derided for doing more damage than good. John Seymour, the Godfather of self-sufficiency, points out in *The New Complete Book of Self-Sufficiency* that 'next year he (the gardener) will win the battle to keep goats out of his garden. What he forgets is that the goats have 24-hour-a-day to plan to get into his garden, and he doesn't.'

It is an entirely valid point because goats are not only intelligent and cunning, they are also extremely hungry and very determined. However, as Seymour points out, goats may be the only chance to rear a milk-producing animal in a small space, such as a garden. In many ways, goats are perfect for rearing in small areas because they are willing to be tethered and respond to this treatment well. In fact, if you fail to restrain a goat and it gets to where it is not wanted, it will wreak havoc.

Continuing on the theme of the goat's capacity for damage, they are also extremely agile, being experts at climbing, reaching, jumping and getting themselves into near-impossible positions to forage. In hot, dry regions of the world where vegetation is in short supply, goats are invaluable and the indigenous people know how to manage goats far better than we do.

REARING AND HANDLING

Goats are companionable. They like company, not only from other goats and animals such as sheep, but also from humans. They are also good herd animals. To this extent, they can be domesticated into the family and, if they are handled from a very young age, you will temper their wild instincts. The best breed of goat for making into a family pet is the Saanen, Swiss in origin, and calm and placid by nature.

If you intend to keep goats, start with a young kid rather than buying in a pair of adult goats for breeding. Make sure you get a nanny kid rather than a billy for breeding and milking purposes. From the moment the kid arrives, it is critical to start to handling it as often as possible. Fit the kid with a head collar as you would a foal or a calf. This will allow you to walk the animal and to train it. The kid will soon learn to seek out and enjoy your company and the collar will also start the process of familiarisation with tethering. Handling at an early age will also help hugely when it comes to milking.

Right: Fencing goats in is critical, as they will find a way to escape given any opportunity.

SHELTER

This is important as it is for all the ruminants, especially during the winter months. Goats do not like cold and wet weather together and need to be brought inside at night to some form of shelter with a roof and dry bedding. In summer, they can stay out.

FEEDING

Goats are not fussy feeders and thrive on a far more varied diet than most animals. They are excellent for clearing ground and will take on the toughest of perennial weeds, including brambles, manuring the ground as they go. If you plan to use a goat for ground clearance, tether it to a long rope attached to a stake driven deeply into the ground. (An iron pin is good for this because it goes in deeply but does not leave a gaping hole on removal). On no account let a goat loose in a patch that needs clearing if there is any possible way for it to escape. For goats, the grass is always greener on the other side. If you are using a milking nanny for land-clearing purposes, supplement her diet with grain and fresh grass or hay.

THE PRODUCTS

An enormously useful animal, the goat has no end of yields. They are great fun to have around, continually arousing controversy and excitement, for they are colourful and character-filled. If you keep animals, it is so important that they are fun. Rearing animals solely for meat, milk or other products can be a soulless exercise. Get the best out of them by getting to know them. Yes, it can be difficult to give them up to slaughter when you have built up a relationship, but if you see your connection with the animal as more than simply the end product, it will ease the pain of separation.

Certainly, goats produce excellent meat. Young kid (around three months old) is as good as lamb any day – fatty, sweet and rich, with barely any of the strong aroma we associate with mature goat meat. Meat from adult goats, particularly male goats, is less palatable, but goat's fat if rendered, stored and used for cooking is extremely tasty and very rich. If you are slaughtering a mature goat, this is undoubtedly one of the finer by-products.

In a small domestic situation, most goats are kept for milk and cheese. Goats yield well, especially when they are happy and well fed. The milk is rich and fatty and although not suitable for babies, on account of being low in

Right: Sheep are very versatile animals in a variety of environments, depending on the breed.

vitamin B12 and folic acid, it makes an excellent alternative for those with allergies to cow's milk. With the backlash against cow's milk, the goat has been placed on a pedestal and the goat-product industry is a burgeoning one.

SHEEP

One or two sheep are as easy to keep in a small area, even in an urban environment. Rather like goats, they can be raised as family pets if they are bottle fed as lambs and handled regularly. They are far less mercurial than goats, domesticating quickly if conditioned to do so. If I had to choose one easy to manage and low-maintenance quadruped to rear in a small garden, it would be a lamb before a goat or a pig.

Unlike goats, sheep cannot be tethered and need to be allowed to run free. Given proper feeding, they will fatten up well without seeing too much grass. Only when a lamb is allowed to develop to three or four months old will the meat hold true flavour. 'Spring' lambs killed at a ridiculously young age barely assault the taste buds. Mutton, on the other hand, is the real meat product from sheep – matured to rich perfection, and there is much more of it.

Lambs are boisterous at a young age and, like goats, they need to be fenced in properly. Domesticated lambs will not necessarily need housing at night as they are fairly tough, except in fiercely cold and wet weather. They will be happy to share a shelter with a goat.

It sounds beyond the realms of possibility, but raising small quantities of livestock in a small urban space is a very real option. Lambs are raised in backyards across the globe without ever seeing a blade of grass. Naturally, they would love to gambol in an orchard with the poultry and geese, but this is the utopian dream – we must make the best use of whatever space we do have.

Farmers and smallholders everywhere raise orphan lambs in the house. Often they are brought in freezing off the hillsides and nurtured back to life in the warmth. Before you know it, the wretched creature is happy to be bottle-fed and to snuggle up to the dog for warmth in its basket.

We must try to avoid the myth that rolling acres of luscious grass are required to raise animals humanely. Unless you are in it for profit, one lamb provides a lot of meat and an excellent fleece. All that is required is that you understand you have an animal in your care. It needs feeding, watering, love and attention to detail on account of the problems that are likely to befall it.

As likely as not, a single lamb is safer in a domestic environment close to the house and humans than it is in an orchard or field. Consideration must be given to the wiles of the urban fox, for young lambs are vulnerable. This is where the dog and the goose are invaluable as early warning signalmen. Their very presence will ward off the threat.

Let the lamb free range wherever there is grass, for that will be its preferred food source once it is weaned off the bottle. If you wish to enclose it, the lamb can easily be fenced in to a mobile pen with poultry. It is equally happy on hard standing as long as it is able to wander about and have access to a food source, such as hay and a supply of fresh water. Lambs have the ability to fit in wherever and with whatever is going on.

The ideal programme for an orphaned spring born lamb would be to raise it close to home through the first year and slaughter it for mutton the following autumn after it has been fattened on rich summer grass and supplementary feed. Keep an eye out for fly strike in the fleece and for foot rot if the animal is grazing wet ground. Assuming good health and well-being, the result will be rich meat and a handsome fleece, especially if you choose one of the rarer breeds such as Soay or, in particular, Jacob, the meat of which is without parallel and much sought-after. Whether or not you will be able to come by an orphaned lamb from a Jacob ewe is debatable, and you may have to settle for whatever needs a home.

left: The yields from sheep are numerous: from meat and milk, to fleeces. In many ways, it is 'the complete animal'.

PIGS

Gardening and rearing animals takes place in all weathers and all seasons and requires a keenness to be involved in the great outdoors. Cattle are good at churning up the ground when they gather around gates and shelter under trees in wet weather, but the pig will outdo them in its ability to create mud. Pigs make a phenomenal mess when they are free to roam. This can be a great advantage for ploughing up and manuring previously uncultivated land. Their natural desire to root around with their trotters and noses will free the ground of perennial weeds and will churn up ground like no other animal.

One other thing before we begin to extol the virtues of pigs: they produce a rich aroma that some may find unpleasant. It is not that they are dirty; on the contrary, they are possibly the cleanest animals of all, but they create a good old-fashioned stink. Finally they are noisy. They squeal and romp like mad, they bellow for food and generally let you know they are around. Apart from all this, they are friendly, intelligent and they love interaction with humans. As well as being engaging, they produce wonderful meat, which can be eaten or processed and stored in any amount of ways, from sausages and bacon to salami and ham. A dream for meat eaters and as far as pound-for-pound value on the holding goes, they are probably the best animal available.

SPACE

Perhaps their best asset is that pigs do not need a great deal of space. Neither do they crave exercise. They are happy padding about in the muck of the pig pen, and providing they are well fed they will produce the goods, albeit slightly more fatty if they are confined. They are not, however, for the squeamish. Whether you choose to kill animals at home, or not, if ever there was one to take to the slaughterhouse it is the pig. Whilst killing a pig at home saves time and money and guarantees that every last scrap of the animal is utilised, it is by no means a pleasant task.

FOIBLES

Pigs will eat almost anything you throw at them. Any leftover foodstuffs from the kitchen will quickly be devoured. Pigs are far less selective than poultry, and they will also eat the kitchen waste that cannot go to the compost heap because it will attract rats. (In the UK, however, it is important to note that

Right: When it comes to entertainment on the holding, the pig is guaranteed to deliver.

'As well as being engaging, they produce wonderful meat, which can be eaten or processed and stored in any amount of ways, from sausages and bacon to salami and ham. A dream for meat eaters and as far as pound-for-pound value on the holding goes, they are probably the best animal available.'

according to DEFRA – the Department for Environment Food and Rural Affairs – you are not allowed to feed any kitchen waste to pigs or poultry if you want to sell the meat commercially.) Any spoilt garden crops, such as windfall apples beyond use or leaf crops which have gone to seed, such as spinach or lettuce plants, will be hungrily devoured. Tubers are a great favourite, but potatoes should be partly cooked to make them easier to digest, although Jerusalem artichokes are acceptable raw.

As robust as pigs appear, they are, in fact, sensitive creatures. They have thin skins and do not like the cold. Place the pig pen out of draughts and shelter it as much as possible. Warm, deep bedding that stays dry is essential. Pigs are clean and will not foul their bedding where they sleep. They reserve their toilet for outside areas. Ideally, a large partly covered pen is what they like, with access to the outside in the summer where they can dust-bath like poultry, scratch their backs on a post and also wallow in mud. If they cannot be allowed outside, provision for such activities must be made inside the pen. Keep the pen clean and remove soiled straw and old, clean bedding either to the muck heap or the compost heap where it will make the best possible manure for the garden when well rotted down.

Similarly their thin skins – and this applies particularly to white varieties of pig – can make them susceptible to sunburn in hot weather. If they are free-ranging outside, they need shelter from the sun. They love to sleep in the cool of the afternoon on a hot day. Be sure to provide a plentiful supply of water in all weather conditions, as each pig will drink two buckets of water per day.

With all this in mind, you need to know how you are going to manage your pigs. Pigs like company. The simplest way is to buy in a couple of weaners (8-week-old piglets) in spring and raise them together for seven months until they weigh approx 55kg (121lbs) to produce pork, or for nine months until they weigh approx 80kg (176lbs) to produce ham and bacon. The pigs will also need good fencing, as pigs are clever and enjoy trying to escape.

Breeding sows open up a whole new project, and you need experience of rearing pigs before you take on this challenge. Sows can be very dangerous when they are protecting their litter. You also need a 'farrowing rail' to prevent them from squashing their young by accident. Litters are large, and all of a sudden you will have a lot of stock on your hands.

THE COW

Poultry, geese, goats, a couple of pigs and even a sheep are all possible to rear on small parcels of land with a little inventiveness and some determination. The 'house cow', the primary purpose of which is to produce milk and dairy products such as cream, cheese butter, yoghurt and ice-cream, presents a sterner challenge. First and foremost, you will need land for grazing on account of the cow's requirement for plenty of green grass.

Does this mean that without land on which to graze the cow it is impossible to raise one? With necessity as the mother of invention, the answer is no. In his autobiography *The Green Fool*, first published in 1938, the great Irish poet Patrick Kavanagh describes how his family acquired a cow in the years preceding the First World War but had nowhere to graze her. His mother would take the cow to the roadside verges 'during the early hours of a summer morning and after rain.' It ended in tears when they were served with a summons for illegal grazing and Kavanagh's mother was fined half a crown.

Times have barely changed since then, for it is unlikely that you could tether your cow on the village green. But that is not to say there are not other ways. As I have stressed throughout the book, the best possible policy to adopt in a 'sufficient' life centred around growing and raising food is to get to know your area and the people in it. There will be a local farmer or landowner who will be able to rent you, at the very least, a corner of a field that you could fence off and divide up so that the grass can be grazed by your cow in rotation. In the urban environment, raising a cow must be a long shot, space being at such a premium, but even on the fringes of a town or village you will find somewhere to graze your animal if you persist.

THE TIME FACTOR

Looking after a milking cow is a full-time commitment. Unlike pigs, geese and goats, which are easily looked after by a neighbour in your absence, milking has to done twice a day, and there can be no misses. If the cow is grazing outside, she has to be brought in, milked, fed and let out again. The milk then has to be processed immediately, especially for cream and cheese-making.

Many commercial milking cows are kept indoors all-year-round and never see a blade of grass or even the light of day. I cannot recommend indoor rearing. This is surely no way to keep a hardy animal who wants to be outside.

Not only do you need an enormous quantity of hay and other feed stuffs, the chances of your cow suffering from a host of different ailments are much higher than they would be if she were to spend the majority of the year on pasture. In winter, it is different and cows are often brought into the cow shed. Grass quality and quantity fall away, and if you feed her outside she will churn up the ground when the weather turns wet. There is nothing she will enjoy more than spending a portion of the winter indoors; turn her out for a few hours a day when the weather is clement and return to her shed at night.

HOW TO GO ABOUT IT

The advent of BSE, the horrors surrounding Foot and Mouth disease and the presence of Bovine Tuberculosis have led to tighter controls on the sale and movement of cattle. If you are considering keeping a cow, you need to know and keep up to date with the rules from DEFRA (Department for Environment Food and Rural Affairs) because they are constantly changing. Channel Island breeds, such as Jersey and Guernsey cows, are unquestionably best for the smallholder. Placid, friendly and yielders of the highest quality milk, they are practically no trouble and very reliable. Guernseys are slightly larger than the very familiar Jersey but are an equally good choice. Like all other animals in your care, it is important to handle your young cow as much as possible. They can then be led to and from pasture and will remain quiet during milking, assuming they have adequate quantities of food at hand. If you are buying a mature cow, arrange a veterinary inspection and check her temperament.

Pests, diseases and health problems aside, the main concerns with keeping a cow are how to feed it and what to do with the product. You should have little problem using up the plentiful supply of manure that she produces when inside over the winter and bedded on straw, but the milk needs a home. Also, your cow will not produce milk each year if she does not calve.

So, what to do with all the milk? A yield of three or so gallons a day (13.5 litres) is a considerable amount. If half of that goes to the cow's calf, you will still have plenty. But be sure of one thing – once you have converted your friends and neighbours to the joys of unpasteurised milk you will be fighting them off, in between making clotted cream, which they will also relish.

Whilst making no claim to be a nutritionist, I fear that the frantic attempts of the authorities to insist on pasteurisation and homogenisation of

milk do little except ruin the quality of the product. Louis Pasteur was rightly hailed a saviour, but those were the days when even basic forms of hygiene were diabolical and milk easily became infected. It is not so today and it seems more likely that we have become more prone to ill health from overprocessed foods. Milk is a classic case in point, for by the time it has undergone all its various clean-up treatments there is little goodness left. Pasteurisation is like an antibiotic; it kills the good bacteria – nay, the essential bacteria – as well as the bad. What is the point? The fact that unpasteurised milk and its by-products are illegal to sell by law is, in many people's minds, a tragedy. The best alternative, if you have the means, is to get yourself a Jersey cow and practise good hygiene because raw milk leaving the udders of healthy animals normally contains low numbers of germs.

Making Cornish Cream

The Cornish people – and I know because I am one – take their cream seriously. We don't call it clotted cream, we call it 'Cornish cream' and, to prove our fanaticism, we are renowned for spooning liberal helpings of Cornish cream even onto ice-cream. Clotted cream is best made using full fat milk with a high fat content. For flavour, it is best to use milk from cows that have been fed on grass.

To make cream, you begin by putting a quantity of milk (perhaps one gallon which equals roughly 4.5 litres) into a large bowl or bowls and leaving it overnight in a cool place so the cream rises to the surface. The following day, heat the milk gently and scald it by bringing it to a gentle simmer (70°C/158°F) for roughly an hour. Do not boil it. Next, cool it down to about 5–6°C/41°F and a yellow crust will form on top of the cream. The critical thing now is to leave the cream until the 'magic ring' appears. As the crust forms on the surface, it will edge away from the rim of the bowl. In due course, a further crack in the crust will form a ring around the bowl a few centimetres beyond the edge. This is the sign that the cream has thickened and is ready to be separated from the milk. Take off the cream with a slotted spoon to avoid bringing unwanted milk with the cream when it is transferred to bowls. The leftover milk is both skimmed and pasteurised. It has been rendered pretty much flavourless and so is best fed to animals, who love it.

DOMESTIC PETS

There is so much to be said for keeping domestic pets on a smallholding. Cats, if not overfed and mollycoddled, especially neutered toms of an athletic disposition, can be invaluable in reducing the population of mice and possibly even rats.

Both mice and rats can do untold damage to crops in store as well as crops in the field. They find their way into feed stores and seed storage cupboards, only to cause havoc when they find something that appeals to them. The time of greatest risk from rodent attack is during their 'hungry gap' between mid-winter and spring when food in the field is scarce. Humane traps work well, but there is a great deal to be said for keeping a cat.

Rats find their way into compost heaps and poultry runs, especially where there is a chicken house or a heap of material for them to burrow under and hide. Rats will feed on kitchen scraps put on the compost heap, and any spilt grain left out when feeding the hens as well as their eggs and young chicks. The problem with rodents is that their numbers increase very quickly. From one breeding pair, you'll soon have infestation on account of the short gestation period. Dogs – in particular certain breeds of terrier, such as Jack Russell's – are particularly useful for catching rats, as are ferrets.

Horses and ponies inhabit the domestic pet world today rather than the working field as they once did. Whilst working horses are an option for those who want to work the land without expensive and polluting machinery, they are scarce. It is hard to justify keeping equine animals just for pleasure, on account of how time-consuming they are to look after and the amount of grass they eat, but they do produce excellent manure. If you keep them outside in a paddock make sure to pick up the droppings to stop the build up of worms.

THE OTHERS

I have discussed only a selection of animals and birds that it is possible to rear on a small parcel of land. Those which I have left out I consider to be either too much trouble for the return in yields. Turkeys, for example, require specialist knowledge and are difficult to manage. They cannot run with chickens or they will contract disease, and unlike geese, they are quite delicate birds. They will not thrive if they are simply left to run in great flocks in a plain field of grass that is fenced in.

Rabbits are farmed for their meat throughout continental Europe, where they run in cattle houses in winter and outside in grass pens in summer. However, a wild rabbit has a better flavour and is available cheaply from most butchers or game dealers. (Rabbits are most commonly available to buy in winter but, unlike most game, they don't really have a season.) Also the grass that the rabbits would eat is better used for multi-purpose birds, such as geese and poultry.

As water is critical on any holding, aquaculture – the rearing of fish – is another area of food production worth considering. However, it is dependent on a very clean supply of water that is properly oxygenated. You can keep small amounts of fish in a pond alongside ducks as long as the ducks leave the fish alone and outside predators, such as herons, can be deterred. You will not be able to use netting on the pond because the fowl will not then have access to the water so you must be vigilant instead. If predators are regularly disturbed, they will quickly lose interest in the fish and look to feed elsewhere.

Down through the centuries, fish have been kept as a source of food and still are in domestic situations. Smallscale fish farming is taken much more seriously in the USA.

STRETCHING THE LIMITS

The greatest mistake is giving yourself too many animals to look after and worry about. Be selective. This book is about enjoying the 'sufficient' life not total self-sufficiency – your survival does not depend on it.

Certainly, a pond will enable you to rear ducks and possibly fish, and for the purposes of bio-diversity in the garden, a water habitat adds another dimension. But stick to what is possible given the land you have available and, as with the vegetables, try not to overstretch its limits. Ask only what it can give, not what it must. Too many animals crammed into a small space may be difficult to manage well, and may have a negative effect on the land, so choose a system that suits you and one that you are able to attend to properly rather than trying to do everything. Remember: the best is the enemy of the good.

HOW WE LIVE

INTRODUCTION

Although this is a not a book specifically about 'green living', I do advocate the importance of a sense of reverence for nature. Currently, we are in the grip of a frenzied clamouring to protect the 'environment', which includes every solution from organic farming to rainforest protection. For me, food is the most critical subject because of its direct association with human health and the strong connection it has with the health of individual ecosystems. What is most important of all, however, is that we take these steps for ourselves in and around our home. This will have a far greater impact on the greater whole because, if nothing else, it may help influence others. This chapter is concerned with making things work better at home. Low energy light bulbs are one way to conserve energy, solar panels too, but the reasons why we have to work towards a less energy intensive way of living need some explanation.

It is not easy to predict the outcome of global warming for this generation, never mind future ones. Some will tell you that economic crises sparked by hunger and lack of food inevitably lead to war, others predict a bleak future after what is termed 'peak oil' runs out – perhaps this has happened already? Whatever the outcome might be, it is likely that energy from fossil fuels will become more expensive, scarcer and unacceptable. What, if anything, will take its place is anyone's guess. But what is for certain is that we are going to have to settle for less much sooner than many of us think.

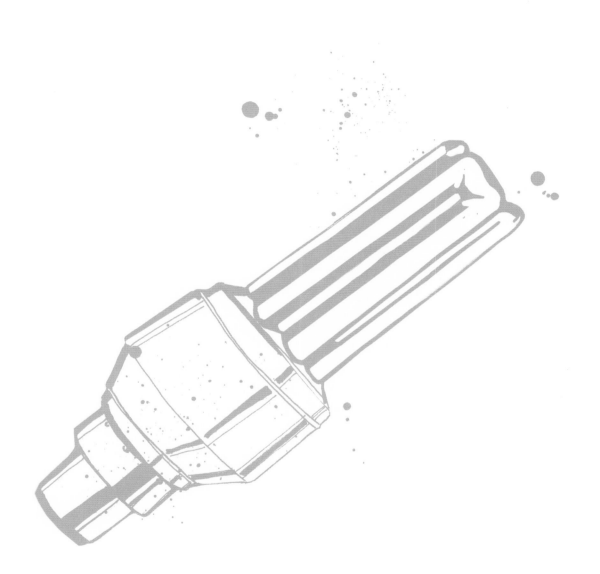

ENERGY

The use of fossil fuels affects almost everyone in the developed world in some way or other. Let's state the obvious just to set the scene: petrol and diesel for vehicles, oil for heating, burning coal to produce electricity. Three big ones. We are not remotely equipped to cope without any of these, and there does not seem to be much of a plan either as the arguments for and against nuclear fuel rumbles on and war destablises the Middle East. Problems, then. Just to make matters worse, the greenhouse gases produced by the chronic over use of fossil fuels and their derivatives are a major cause of global warming. From bad to worse. Just to tie it up, almost everything in our houses, everything that we own is manufactured from some sort of by-product from the petrochemical industry.

Where I live in the south-west of England, an enterprising group of people have come together under the banner of 'Transition Town Totnes' (Totnes, Devon, being the town in question) to try and look at the future for the town and its inhabitants beyond 'peak oil'. This is no disaster action group peopled by prophets of doom; it simply sees the inevitability of a less consumptive lifestyle as an opportunity for great change. That change can only be for the betterment of nature as well as mankind.

As stated on the home page of the group's web site, 'The thinking behind Transition Town Totnes is simply that a town using much less energy and resources than we presently consume could, if properly planned for and designed, be more resilient, more abundant and more pleasurable than the present. TTT strives to be inclusive, imaginative, practical and fun.'

At the centre of the group's focus is an 'Energy Descent Action Plan'. This is a realistic look at life beyond carbon, where 'the best ideas are inspiring, creative and attractive visions of revitalised local economies, visions grounded by a connection to place and the people in it.'

That I live in Totnes has no relevance, but my feeling about such a group and their plans is one of relief. These are sensitive times and what TTT knows instinctively is that in order to attract people's attention and input the organisation, its ideals must be attainable. The key words here are 'inclusive' and 'fun'. For far too long, the Green movement has been bedevilled by self-importance and arrogance. What TTT says is, Come and join us; we are inclusive not exclusive; these issues affect us all and together we can make a

Right: Power is the most pressing issue, both for humans and the environment in which we live.

difference and enjoy ourselves while doing so. The emphasis is very much on permaculture and the practical solutions offered by that system. Anyone who has practised permaculture or been around people who think laterally after that fashion, know that it is great fun, a bit like living outside the law. And this for me is the key that, just like the gardening and the reverence for nature, the entire concept should be fun. This is something for everyone, that should be designed by and for everyone and should be a common goal. If it is not any fun, then it simply won't be attractive and we will all have to struggle on doing our bit on our own.

The TTT's 'Energy Descent Action Plan' points to redesigning our lives so that we reduce our energy consumption on all levels. It is permaculture in action. An example of a country that did this is Cuba. The Cubans took to permaculture on a grand scale after the collapse of the Soviet Union in 1989, when Cuba's oil imports were more than halved at a stroke. Organic farming was embraced as a low energy system and Cuba pulled together because it had too.

All this human energy concentrated in a small area can effect great change and should be encouraged and fostered. A descent plan, as described above, has the community at its heart, and this is where we should be focusing our intentions, as a collective. Take individual action, yes, but as a group collective there is more, well...energy.

INDIVIDUAL ACTION

Clearly, there are many things we can do to reduce energy output directly, starting with turning off the lights when leaving a room, using proper insulation and switching to a 'green' energy supplier. But the 'permaculture mind' thinks a little more laterally about low-energy output and how to make the best use of the energy that we do expend. Let's think back to the garden for a moment. Do you remember that I suggest housing your chickens in the polytunnel over winter to make use of their body heat to keep plants warm? (See page 208). That is the kind of thinking that we need to be doing.

It all starts and finishes with the sun because this will probably be our major energy source in the future. It has already provided us with all our fossil fuels and it will probably have to solve our current energy crisis.

Designing with the sun in mind is important when planning the position of a new house, garden, vegetable plot, or even a window box. We must consider how best to capture its heat, let its heat escape, and so on. Gardening

Right: Solar panels occupy the roof of this house. An emphasis on renewable energy has to be made so we can become more self-reliant.

in shade is difficult, and either side of summer there is little energy value in the heat of the sun, especially in the northern hemisphere and particularly in winter when the sun has a very low trajectory.

The great thing about the sun is that it is free source of energy and there is lots of it. Trees are a very good place to store energy from the sun and convert it into carbon to provide fuel. If you plant trees as a sustainable fuel source – and they are growing faster than they are being cut down – then you will be in 'carbon credit'. Also, trees use carbon dioxide to grow whilst at the same time producing the oxygen we need to breathe. Don't let anyone tell you that trees are not the answer to saving the earth: without green plants – and trees are the biggest of these – we would not have oxygen. All plants hold the sun's energy and turn it into both food and fuel.

A way of conserving energy is to install a turf or living roof, which are well-known for their insulating capacity as well as their ability to keep buildings cool in summer. They also have extra benefits, such as attracting insects, blending in with a rural backdrop, and, slowing rainwater run-off and so reducing the risk of flash flooding. Also, a roof covered in grass, like those on alpine chalets in Switzerland, can reach down to soil level and feed a tethered goat. A roof covered in tiny flowering sedums is one of the most attractive sights. It will not only impress your neighbours but show them that 'appropriate' technology can be beautiful as well as energy conserving.

Solar energy in its less direct forms, via solar cells and panels, can provide you with electricity and hot water. Simple to install and manage, and becoming cheaper by the day, this renewable and attainable source of energy has to be the future. Just think how hot a room can become when the sun is out, even on a winter's day. If that is what passive solar heating can do, then trapping this huge source of energy for redistribution around a house must be an essential course of action. Add to this the fact that the solar cell and panel systems can work on cloudy days, and there would not appear to be a better means of harnessing the sun's energy.

Wind power remains a hugely untapped resource but one that is growing steadily. In the UK, less than a single figure percentage of electricity produced by power from the wind makes it into the national grid, but this will change as renewable sources of power become more widespread. The small contribution made by wind power is not because it is inefficient or because there are planning issues concerning new wind farms; it is simply because we have not

Left: Use of natural resources for energy conservation is becoming more widespread.

got around to realising its potential. Just ask the state of California, where somewhere in the region of 15,000 turbines produce enough electricity to light San Francisco, or Denmark, where 20 per cent of the nation's electricity comes from wind power and their clean energy industry is worth more than 3 billion euros annually, employing more than 20,000 people.

Nonetheless, high ground that is open and exposed is critical for wind capture. From a home-users point of view, there are small domestic wind turbines on the market that can capture enough wind to charge a 12-volt battery or provide power for the entire house. If you choose to go 'off-grid' and provide all your own energy, then you will need back-up in case the wind fails, and this back-up usually takes the form of a diesel-oil fired generator. In an ideal world, this generator could be solar-powered.

Coming up in the ranks as sources of alternative energy are geothermal exchange systems or ground source heat pumps. These, whilst requiring pipework and pumps to move the water around underground, work on the principle that the earth can be used either to heat or cool water. The system, which requires a bore hole to be dug to bring up water, is especially suitable for underfloor heating systems, as the temperature does not reach very high values. It is cheap to set up and entirely renewable with no extra inputs, apart from the cooling liquid and a minimal amount of energy to run the pump.

Wave power turbines are rapidly being developed, with the UK the leader in the field. The US department of energy estimated that over the next 20 years more than one thousand new power plants will be needed to meet the country's growing energy needs. That puts the whole energy debate into perspective.

Right: Deriving energy from wind power is still a subject of debate in some countries, due to the public's objection to the windmills themselves.

An Inconvenient Truth

Whatever anyone thinks about Al Gore's film *An Inconvenient Truth*, it will surely help to bring the issue of global warming into the mainstream.

The former US Vice President (1993–2001), during the Clinton adminstration, and committed environmentalist, has made a film that absolutely everybody must see if only to reinforce what we already know: that the world faces a massive challenge to stop the destruction of the planet, which means that we must all take on moral responsibility for our individual actions.

This is a film that appeals to everyone whose life is dependent on the ways that are destroying the planet. It is quite clear on issues of global warming and the destruction it is causing. Gore is not afraid to make scary future predictions as to what might happen unless we curb our actions.

Nonetheless, Gore is convinced that it is not too late and that we can make a difference collectively if we take on this moral responsibility. I agree with him and have urged similar responsibility, albeit on the subject of growing and producing food. The issues of mass-scale industrial pollution and climate meltdown are beyond the scope of this book, but I do know that every little effort must help the wider cause, and this is what we must concern ourselves with here. The fallout that no one talks about and which Gore fails to mention, is that once reductions start to kick in those that blatantly fail to comply will probably be ostracised from society. What form this will take remains to be seen.

Gore's website www.climateecrisis.net advocates 10 simple things to do to stop global warming and help reduce carbon dioxide use. They are: change to eco light bulbs, drive less, recycle more, check your tyres, use less hot water, avoid products with excess packaging, adjust your thermostat, plant a tree, turn off electric devices and finally, of course, buy *An Inconvenient Truth*.

All laudable, but I would like to make a few additions, the first being the need to develop a reverence for nature. Saving water of all kinds is critical. Making compost and planting a garden of any size is essential, but perhaps the most important of all is to exercise the will and develop a feeling of 'sufficient'.

WATER

After energy comes water. It is as important as energy, and the lack of it is likely to become a massive problem in the future. The simple reason is that we have little control over it. If anything, the sun is likely to increase in its capacity to give energy whereas water in the form of rain may well decrease. When this happens, things will get sticky.

As a race, we are as wasteful of water as we are indiscriminate in our use of fossil fuels. We are completely dependent upon water, not least because 90 per cent of our bodies are made up of water.

As an example of our dependence on water, I recently had a late-night panic when our boiler developed a fault and it had to be turned off. We were without heat and water for eighteen hours. That meant no water for cooking or washing or even brushing our teeth. Also, because the heating system runs on water, we were without central heating.

Not ideal by any means, but life without running water and heating is the normal state of affairs for millions all over the world. My point is that here, in the developed world, we are entirely dependent on a constant supply of running water. The question is how we can come by it without having to rely on an expensive supplier and, what's more, an expensive supplier of a vastly inferior, chemically cleaned product. Recycling is one thing, but the water we drink today tastes disgusting and has been through bodies before. This cannot be right. Yes, every drop of water is ultimately recycled, but in the natural state of affairs what we have used already, comes back to us in the form of rain that has come from the oceans. What is needed is a proper system of purification, free of chemicals, rather than having to drink a product that has come from a purifying water treatment plant via a reservoir, over and over again.

Perhaps moisture holds the key. Unlike water, moisture is a curious commodity because we can create it using natural means. We can't order the sun to come out just as we can't command rain to fall, but we can produce moisture, so water can be realised where none existed before. The trick is how to hold on to that moisture once we have it and how to harvest and make the best of what comes to us in the form of rain, springs, run off, snow and ice – or any other form that the moisture chooses to take. Plants, particularly trees, are very good at producing and retaining moisture. Not only do they produce moisture as they grow but they draw moisture up from the ground and hold it

for release when required for use by themselves and other plants. This is the natural order of things. This is why the forest is the most stable environment on the planet and why forest gardens are being modelled all over the world as a complete system for food, fuel, fibre, fodder, energy and water.

RAINWATER

For most of us, the best way to collect and store rainwater is to install a water tank, as water butts will never hold enough water for all your household and plant-growing needs. Before you buy or build a house, you need to think very carefully about the water supply: whether there is too much and a risk of flooding, or low rainfall and the chance of drought; what the water quality is like, and whether or not you can come by your own supply. Not many of us are lucky enough to move to rural areas, where there may be a plentiful natural supply. If you do move, then access to your own water supply is the first and most important thing to consider when looking at a property. Water will be one of the most sought-after commodities of the future.

The unpredictability of the weather means that we have to be better prepared for drought or flood. In 2006, who would have thought that there would be a ban on the use of hosepipes in the south-east of England as early in the year as February? The cause was five dry winters in succession (winter is the time when the bulk of our annual rainfall occurs), and as such, water tables and reservoirs were extremely low. The hot summer of 2006 that followed opened a lot of eyes to the need to conserve this precious commodity.

The dry spring of 2006 in the UK was followed by that exceptionally hot summer, during which there was virtually no rain at all until the latter quarter of August, which sent people spiralling into panic. It was a feast day for the media, which forecast impending disasters and thankfully kept on about global warming. For many, a prospect that had not been required since the summer of 1976 loomed large: stand pipes in the street.

Harvesting rainwater for use at a later date is essential, and there are many ways it can be used in the garden. Firstly, rainwater is 'soft', and it is soft water rather than the chemically recycled 'hard' water that is favoured by plants. Any simple water collecting and distribution system must allow for water to travel around the garden to be used where needed.

But which plants need watering and how often? Seeds and seedlings planted out into open ground need watering, but once a plant has taken and

Right: All efforts should be made to save as much rainwater as possible.

its root system is established, it should be able to fend for itself, especially if the soil is fertile and moisture retentive. The idea is to encourage plants to put down roots and to look for moisture in the soil rather than making them dependent on watering. This makes for a tougher more resilient plant.

During a long, dry summer, plants can become stressed, and the watering situation must be reviewed (particularly for the smaller more vulnerable plants that root close to the surface of the soil). Small plants may need help as the soil surface dries out, but what is important is that you do not get drawn into a constant round of watering plants that are growing in open ground. For those plants growing in containers, things are different: they need twice-daily watering in hot, dry conditions, as their roots are confined in the pot and so are not free to seek out water.

Perhaps the greatest tool in the garden for saving water is mulch. Whatever material is used – compost, manure, straw, bark chippings, or leaf mould – the result will be more available moisture around the plant roots. (The natural state for most plants is to have leaf litter around their base, by which the roots are kept cool and the moisture is retained.) Whilst it probably will take too much time to mulch a large number of annual plants, you should aim to mulch all your perennials, trees and shrubs, as they will benefit. Mulching strongly reinforces the argument for making as much compost as possible by recycling your household and garden waste. As well as saving water, this will also recycle precious nutrients and put them back into the system.

While a decent-size tank to hold rainwater is essential for the garden and for watering livestock, it will also serve you inside the house. A relatively simple plumbing alteration will see rainwater in the bath, the shower, the washing machine and other household appliances. There is no reason why you should not use it for everything except drinking, and when it becomes 'grey' it can be recycled again to water the garden.

GREY WATER

Collecting rainwater may seem like a logical step forward in the drive to conserve water, but the reuse of 'grey' water is less widespread. 'Grey' water is the term used to describe water that is quite simply dirty, whether it has come from the shower or bath, kitchen sink or washing machines. 'Grey' water can be used directly on the garden without too much worry as long as it is not too 'detergent heavy' and the detergents you have used are of the ecological kind.

Grey Water

Whilst it takes some commitment to use a compost loo and recycle human waste, it is relatively simple to recycle 'grey water', which is household water that has been used but has not passed through the body. Grey water that has washed dishes, clothes and bodies in baths or showers can be recycled to water plants or be used to flush loos.

The northern regions of Europe have, up to now, been well off in terms of water. This is no longer the case. The last five winters across mainland Europe and the UK have been noticeably drier, and we are running short. In Mediterranean regions, where rainfall is scarce and unreliable and water is overused, drought has become a very serious problem.

Like the permaculturalists, I have always believed that plants should be encouraged to send their roots down to look for water. This also allows nutrients to be brought up from the lower levels of the soil and makes for robust plants. However, if drought conditions prevail, watering becomes a necessity. For this purpose, grey water can be held in a butt or tank and used when required.

I first used grey water in the drought summer of 1976, the first time that households in the UK had to learn to deal with drought conditions. My approach was rather amateur. I took the hosepipe up to the bath, flung one end out of the window into the garden and sucked on the hose to create a vacuum and draw the bath water down the hose. This system watered the lawn and the runner beans.

Grey water use is limited but, if it is not heavily-laden with detergent, it is safe to use on edible crops. I do this by laying a leaky hosepipe system from the water butt through the vegetable garden. I direct the water at crops that are known to be thirsty (particularly beans). To retain water in the soil and prevent evaporation, I also place a mulch on top of the ground after watering. In a summer of drought ,where long hours of sunshine further dehydrate soil that is already dry, evaporation happens all too quickly. Cover that area with a mulch after watering (tuck the leaky hose underneath the mulch), and evaporation rates will be reduced. A biodegradable mulch will keep weeds at bay and add nutrition to the soil. Straw is best, as it will not block up the tiny holes in the leaky pipe.

Clearly it is not advisable to use such 'grey' water on edible crops, such as salads or young seedlings, but trees and shrubs and any grass will welcome it in hot weather. Simple solutions also exist, such as the reed bed system described below, whereby 'grey' water can be cleaned up and used again.

REED BED SYSTEMS

One step on from recycling 'grey' water is recycling sewage as it uses huge quantities of water to carry away human effluent. Most of our sewage is pumped to a treatment farm where it is cleaned up and either put back into the system as 'safe' water or expelled into the sea. A reed bed system to treat sewage has clean water as the end product as well as reeds, which can be harvested and used for compost. Reed bed systems are becoming more common and an increasingly used method that is effective, safe and relatively cheap. It is a very good system for small communities, where the water can be recycled and put to good use. Space is the limiting factor because an area approximately 30m x 10m is needed for a reedbeed system to service a family of six. (The waste is flowed through a series of different settling tanks, and gravity-fed oxygenated water flow forms and space is required for the reservoir itself.) The treatment can be done undercover, but the reeds need access to light in order to grow.

The authorities that are responsible for disposing of our sewage do a fantastic job and we have a very advanced system of handling sewage and recycling it – if, that is, you are happy with chemical treatment to clean up a substance hazardous to health before it is put back into the system. A reed bed has absolutely no side effects and is completely safe. There is no sludge and no smell, only reeds and water. Gravity accounts for the flow of the sewage through the bed and there is no machinery required. Every settlement should deal with its effluent in this way.

Right: Reedbed systems for sewage treatment save huge quantities of water, as well as producing a useable by-product in the form of reeds.

COMMUNITY

The trouble with being green is that it still, after all these years, has a whiff of worthiness about it. People are cagey about the whole community thing. I've been advocating organics for 20 years or more and still it has its detractors. There is a stigma attached to an 'environmental' approach to life that catches in our throats and somehow stops us from taking hold of the ideals behind it, worthwhile or not. It's going to have to change because the government is now committed to emission reductions. There is no choice. All of a sudden, we are going to have to act greener. How will this make us think and how will our approach to life change? It's amusing because surely no one ever thought this day would dawn when we'd have no choice but to change our way of living.

How are we to prepare for all this change? It is only since the 1960s, after the publication of Rachel Carson's *Silent Spring*, that environmentalism began to take hold. That book highlighted the damage done to the planet by a reckless use of chemicals and an utter disregard for nature. Thankfully, we now know where we are with all this – on the brink, looking over the precipice. Global warming is in full swing, the planet's soaring populations are completely overburdening their resources, there can be no possible means to balance this state of affairs. As I have kept on urging, if we took more notice of the glory of nature we might perhaps find the will to stop destroying it. In his book about communal living, *Utopian Dreams: a Search for a Better Life*, Tobias Jones travels to different communities in Europe. From his experience of community living, he believes that perhaps the only thing that unites us as a race is our desire to be different and our desperation to promote the rights of the individual. Jones still concludes, however, that under a banner of working for the common good as a group, life has real worth and is far less lonely than it is if we go it alone.

I think that we have lost our way both morally and spiritually. What a sense of community and community living is concerned with is taking responsibility for our own actions for the benefit of the greater good. Two of the great religions of the world, Christianity and Islam, have this belief at their core. Whilst Christianity is in decline in the West, it is growing in Asia and Africa, as is Islam. On these Continents, the Church encourages its followers to practise this strong sense of moral responsibility. Yet few dare to point this out, just as they seem unable to drop their desire to act in their own interests.

In 2006, Jonathon Porritt, the Former UK Chairman of the Green Party and Friends of the Earth, charted the history of the magazine *Resurgence*, which for 40 years has championed spirituality and the environment. *Resurgence* he said, 'has become the voice of those who not only believe that today's crisis is characterised as a crisis of the human spirit, but who are convinced that much of the leadership and inspiration that we will need to navigate the road ahead will be more spiritual and moral than political and economic'. This must be close to the truth. *Resurgence* mixes spirituality with environmentalism, the arts and forward thinking for the modern age with a strong moral message: 'that the future of mankind and the planet that he inhabits is down to individual responsibility'.

Community, as a theme and a practice, encourages the development of individual responsibility, and as Jones points out it allows people to 'congregate responsibilities' for the betterment of all concerned within the group and in the wider world. Communities, based on shared beliefs, have been around forever and start, of course, with the family. 'Alternative' communities have a role to play as guiding lights in an increasingly muddled world.

INDIA

In the 1980s – not long after completing a two-week permaculture design course – I set off for South India to help on a community project. At the time I was studying commercial horticulture and, as part of my course, I took one year out to go and gain some practical experience. My fellow students went to work in tomato-growing units and cut-flower production plants, while I ended up trying to keep wild pigs out of the rice crop in a picturesque valley in Tamil Nadu. Here, I was working on a mixed forest farm, where coffee, tree crops, annuals and livestock were in the system. The founder had also established a school, ran a homeopathic clinic for local people and revitalised what had become a very rundown estate. My interest in the project came from a desire to learn as much as possible about sustainable agriculture and localised development on the Indian Subcontinent.

The farm was run organically, and 'appropriate' technology was incorporated in the shape of solar and wind energy, plus 'biogas', which was produced by the cattle who were housed at night. It was a fantastic introduction to what could be achieved at a local level. The investment on the part of the owner/founder was considerable, but the benefits for the local

people, the environment and the area was there for all to see. The sense of community and purpose behind each action was also clearly evident. For me, the sun, the delicious local wild rice, and the fresh mangoes and avocados the size of footballs picked straight from the forest trees, were beyond the wildest dreams of a young horticultural student. Whilst nature's bounty made the working days sail by, this was no utopian commune. It was a project that focused on development of the community and the mind, and whilst it was beset with problems and had a pyramid-shaped management structure, its intentions were honest and far-reaching. For me, it worked and I could see that it also worked for others, particularly the local farmers, who were more used to the unsustainable ways of 'slash-and-burn' farming. The area became revitalised.

My stay in Tamil Nadu also led me to one of the outstanding communities in existence anywhere on earth, Auroville. It was established in 1968 as a world city of peace with the desire to be a 'universal town where men and women of all countries are able to live in peace and progressive harmony above all creeds, all politics and all nationalities'. It has some 1,700 members and has achieved virtual self-sufficiency in power and food.

Tight to the east coast of south India, where the heat and humidity are overwhelming, Auroville was established by a French woman known simply as 'The Mother', who worked with the Indian sage Sri Aurobindo. It is a community based on strong spiritual and environmental influences which bring it together. In his book *Ecovillages, New Frontiers for Sustainability* Jonathan Dawson traces Auroville's success to three core factors. Firstly, a 'strong spiritual impulse that acts as a powerful community glue. Secondly, its status as a symbol of hope and unity has afforded both friends and money from within India and internationally, and thirdly the community has developed a strong economy of solidarity that has served to redistribute financial wealth within the community.'

It is a tough place, Auroville. The climate is inhospitable because of the heat and a lengthy wet season, but the common purpose has led to inner strength, and herein lies the key. 'We are all friends until it comes to money', as the saying goes, but somehow the Aurovillians have managed to engender success because they believe in a common bond and have a strong spiritual and moral belief system.

ECOVILLAGES

If nothing else, communities like Auroville or Findhorn – a famous spiritual community on the east coast of northern Scotland – (see sidebar), are excellent places to look for inspiration about how we might garner our resources or even gird our loins for the future. Most of us live in towns, and more of us will soon. The Chinese are building 'eco-cities' in recognition of this, and eco-communities are being established in the UK, such as the Beddington Zero Energy Development project (Bed ZED). More and more people not only want to realise the benefits that are afforded by sustainable community living but are going to have to as physical and moral resources dwindle.

Dawson himself is president of the Global Eco-Village Network and lives in the eco-village at Findhorn. He practices what he preaches, and his book is a revelation. In summation of this extraordinarily powerful young movement, his enthusiasm and passion are hard to ignore: he talks of the sheer energy and creativity within the GEN movement. 'Ecovillages' he says 'are pioneering new models on multiple fronts. One is struck by how often they are in the vanguard in introducing new techniques or models – organic agriculture, community-supported farms, building techniques, mixed special needs and non-special needs groups, community currencies, solar technologies, biological waste water plants and so on'. However, he rightly points out that although the small scale and shared values of eco-villages serve them well, 'there are many challenges that face us today on the path to sustainability that cannot be addressed at the level of the individual, or even small group and for which community-level action is required'.

AT A LOCAL LEVEL

Buying a house in an eco-village or joining a community is beyond most of us, but it is from these paragons of 'ecotopian' virtue that much can be learned. We might be inspired to build a house from straw bales or put in a sustainable water treatment system at the bottom of the garden. But what about those of us who live in towns? How are we to feel 'sufficient' without the support of other like-minded members of such communities? How are we to realise the sense of freedom that comes from lots of available green space? The answer is staring us in the face: we already live in communities, which is what towns are – tailor-made with shared communal spaces such as libraries, shops, car parks

FINDHORN

The spiritual community at Findhorn in Scotland has been a source of enormous inspiration for me and many others since I first heard about it in the early 1980s. I had just begun to grow organic vegetables properly and take an interest in all things environmental when suddenly I came by a book about the garden at Findhorn. It told of giant cabbages being grown in the pure sand of the land around the Findhorn estuary with help from 'nature spirits' as opposed to organic matter. I vowed to go there to see for myself, and though it was not until twenty years later that I signed up for 'Experience Week', it was worth the wait.

What I found at Findhorn was a vibrant community encompassing all the elements that may be able to help us through the modern era. Strong spiritual values, home-grown organic food, waste-water treatment, community-supported agriculture, eco-housing, and people – like-minded people who run the place themselves, finance it and work in it.

What began as the vision of three people, Peter and Eileen Caddy and Dorothy Maclean in the early 1960s has become one ▶

and entertainment venues, plus lots of the raw material that is the bedrock of all communities: people. Today, facilities and communications exist to make life easier and more comfortable than it has ever been. We have only to make the choice to car-share or lobby for more green space or get an allotment or use an ecologically sound builder to make a difference.

It does not matter where you live or how you live, but in order to get the best out of your community it is important that you put your best in. We have to learn to co-operate with each other more than ever before, and one of the ways to do that is to understand the place where you live and to become an active part of it.

The UK-based organisation Common Ground is one of the best-known supporters of what it calls 'local distinctiveness'. They seek 'a link between nature and culture and focus on the positive investment people can make in their own localities'. They have done much to support schemes to put back what are called 'community orchards' in recognition of the fruit orchards that we have lost as more and more green spaces are developed.

Trade is also vital to the health of a community. Not only does most of the money exchanged stay within the community, but there is interaction between those who are trading. An exchange of something as simple as a greeting can mean a lot in this disparate age. It doesn't take much effort to join in and feel part of your community.

Not long ago, one of our cats went missing. I put a notice in the Post Office. Two days later, I got a call. The lady had found the cat, deceased, scraped him up and taken him home until the flies had got him, when she took him to the vet, who put him in the deep freeze. She saw my note and rang up. It was all way beyond the call of duty and she was not specifically a cat-lover, but that's community for you!

RUDOLF STEINER

Born in Austria in 1862, Steiner founded the Anthroposophical movement and was responsible for the Waldorf form of education. He was also the driving force behind what became known as bio-dynamic farming.

Steiner's lectures on agriculture, from which the bio-dynamic movement emerged, were delivered to a group of farmers in Poland 1924. They were the result of grave concerns over the failing health of soil and plants and the high incidences of pest and diseases in the face of widespread chemical use.

of the most successful communities in existence in the northern world. Run as a charity with more than 300 members on the sites at Findhorn and Cluny and countless other worldwide, the work of the Findhorn Foundation reaches far into the community as a whole. No collection of idealistic hippies, these. Hard-working, centred and committed, the members of the community have made this an extraordinary place.

Community living may not be for everyone, but there is a sense of purpose and a spirit about the members of Findhorn. They eat and work together and run the place with and for each other.

Headquarters of the Global Eco-Village Network, Findhorn is also the base for the charity Trees For Life. This is an organisation that is steadily going about one of the most prolific reforestation programmes anywhere on earth. Its aim is to replant the Great Caledonian Forest, which once ranged over most of Scotland. Today, against all odds, they have planted more than 500,000 trees.

Left: BedZED eco-village in the UK. Urban communities present huge potential for energy saving, where the most precious resource of all is present in quantity: people.

In the lectures, Steiner put forward the idea that a spiritual dimension had to be recognised in all approaches to nature and in particular agriculture. Whilst bio-dynamics is founded on good husbandry, it is what we call organic today in that it is free from chemical use. Steiner's system requires that the farmer or grower view the earth as a living being, one that is open and receptive to all the influences that stream in from the cosmos, where the spirit is present in all living things.

The degradation in nature today is a reflection of man's relationship to this spiritual world. Thankfully, a strong ecological movement argues convincingly that the universe is not a machine but rather a harmonious whole made up of different dynamic relationships between organisms.

The bio-dynamic movement's approach to farming is that the farm, like the universe, should be seen as a single, whole organism and, as is the opinion of many permaculturists today, should not be dependent on outside inputs.

The method of farming that Steiner encouraged from those first lectures has developed into a worldwide farming movement or system, based on organic cultivation, which has its own certification system (Demeter). Beyond organic principles, bio-dynamics works, amongst other things, in relation to the cycles of the moon and the constellations. If we stop for a moment to look at the influence of the moon on three areas of life – tidal movement, menstruation and lactation – we can immediately see just how strong that influence is.

Deeper into the practice comes the requirement to use certain 'dynamic' plants to activate compost heaps and to make preparations with which to treat those compost heaps and the soil itself. That aside, Steiner was clear about one thing: the practise of what has become known as bio-dynamics could only work alongside good husbandry. His aim through bio-dynamics was to bridge the gap between the terrestrial and then spiritual.

WALDORF EDUCATION

One of the largest independent education movements, Steiner's Waldorf form of teaching (named after the place where it was first taught) encourages an imaginative approach to learning through the development of creative as well as analytic thought. The education's ultimate goal is to give young people the basis with which to develop into free, moral and integrated individuals.

Right: Steiner education places great emphasis on interacting with nature from a young age. The Steiner School in south Devon, UK, which our daughter attends, has its own bio-dynamic market garden, run with huge input from the children.

From kindergarten age, learning happens through imitation and example., and from story telling and creative songs, poems and games. There is a lot of free play and great emphasis is put upon reverence for and interaction with nature. A lot of time is spent out of doors, gardening, walking, playing and creating. The school at which my daughter is a pupil has a large bio-dynamically run market garden, where produce is grown and sold by the children.

As a Christian, Steiner put strong emphasis on the seasons for both their physical and spiritual roles. Thus not only are Christmas and Easter recognised as the two most important festivals in the Christian calendar, but Michaelmas, Martinmas and the Saint's Days are also celebrated. The rhythms of the year are recognised through these feasts.

Beyond kindergarten, the teaching becomes more arts-based with visual arts, drama, movement and music coming to the fore. Languages are also introduced from the ages of seven onwards. By the time secondary school age, 14, is reached, the curriculum becomes much more academic, but the vital life components of responsibility and purpose are also taught. The atmosphere is non-competitive and individuals are encouraged to develop at their own pace and with a strong eye towards ethical principles and social responsibility. This is a wonderful form of education that functions around a very strong community of like-minded people.

TREES FOR HEALTH

You can be absolutely sure that somewhere near you, whether it is via a large organisation such as The National Trust or even your local authority, there exists the means to get out in the countryside and get involved. For those of us without gardens or access to green space, this is a valuable resource.

Wildlife Trusts, animal and bird groups, restoration projects, volunteer outfits – there are no end of such organisations to enable you to appreciate and work with nature. Much better and more fun to go local; it will mean much more to you. In my very early days of gardening, I used to 'WWOOF' – the acronym stands for 'willing workers on organic farms'. It is a volunteer organisation that places workers on organic farms or holdings for as little as a weekend or as much as an entire season. If organics are your thing, then this is a great way to gain some useful experience.

Trees For Health is another group that works to 'reconnect people and plants' in South Devon. Its aims are to benefit the health of landscapes

Left: The forest is the most stable environment on the planet; what better place to discover a reverence for nature?

Robert Hart

I have discussed forest farming as widely practised throughout Asia under 'homestead gardening' (see page 63). However, we owe its development in the first world to the late Robert Hart. In the 1980s I had the privilege of making two visits to Robert's forest garden in Shropshire, UK. The total area of this extraordinary forest garden was no more than ⅛ acre (0.05 hectares), tiny, but from it Robert produced almost all of his food for eight or so months of the year. It was a haven and an example of what could be achieved in a small area. More than that, it was a testament to how systems that model themselves on nature can work.

The first time I came into contact with Robert Hart was through the book *Forest Farming,* which he co-authored with James Sholto Douglas on agroforestry. It was a fascinating book that explained how trees could play a huge role in not only stabilising the planet but also produce enormous quantities of food, animal fodder, firewood and so on.

In true permacultural fashion – although he had been practising permaculture long before the phrase was coined – Robert Hart was concerned with allowing nature to produce its bounty with minimal interference. In a similar fashion to the natural farmer, Masanobu Fukuoka (see page 24), Hart knew that perennial plants had the ability to source nutrients by virtue of their deep roots and thus had no need of fertilising. He also knew that they would go down to find moisture rather than depending on an external water supply.

Robert's was a non-interference garden based on the 'seven storeys' of the forest. I will never forget sitting in Robert's kitchen discussing what was then a revolutionary method of farming and growing in the northern hemisphere. Surrounded by drying fruit and herbs from the garden, he gave me a thorough grounding in what it meant to do what he did. His garden was overflowing with fantastic fruits and perennial herbs and vegetables, including one of the finest Japanese wineberries I have ever seen. A finer exponent of the art of the forest garden there has never been and everyone should read his book *Forest Gardening*.

and people as well as revitalising the food and medicinal use and knowledge of woodland species. Trees for Health draws our attention to the important fact that, although our British landscape was once forested, woodland currently accounts for just 8 per cent of the total land area. Although most of this woodland is in private ownership, Trees for Health works with these landowners to gain access for educational purposes.

Such an organisation is an excellent place to learn about trees without having to attend a specific course or college. Activities are numerous and include tree planting, seed collecting, propagation and harvesting. They also work with primary schools in the local area.

Trees for Health point out that our connection with trees has always been an integral part of our culture and that we have forgotten this. Just seeing and being amongst trees has direct health benefits and this simple experience can reduce stress levels and blood pressure.

Trees provide the land with a 'skin', a protective layer, and they are quite literally the 'lungs' of the planet. This is why we make such an issue about protecting the rain forest. Let's be quite clear, your little patch of community woodland is as important as the Amazonian rain forest. It prevents erosion, holds countless species of plants, animals and microbes and reduces the impact of climate change. Supporting the work of such organisations as Trees for Health has immediate and long-lasting benefits for trees and the community.

RECYCLING

Recycyling water leads us onto the subject of recycling all sorts of waste materials. At the top of the pyramid for recycling is human effluent (as previously discussed in the Reed Bed System, page 256), so let's start with this. As with other organic matter, if human effluent is properly treated, you can remove the pathogens and odour to produce compost for the garden. For four thousand-odd years, effluent has been carried away by water, and since the nineteenth century the water closet has become commonplace throughout the western world. The use of 'night soil' – the contents of cesspools removed at night, especially for use as manure – has also been in use for thousands of years and was important to the Chinese. Theirs was a cyclical 'closed-nutrient' system, which means that the nutrients in the 'night soil' went to feed the crop, which in turn fed people and then went back into the crop. (Interestingly, Kolkata [formerly Calcutta] in eastern India grows all of its vegetables on its sewage site just to the north east of the city from some eleven million gallons [50 million litres] of raw sewage water, which are channelled there every day.)

If you are ready to take the step towards recycling human effluent, there are several 'compost loo' systems available. There are systems which can be used in a bathroom with tanks, smell traps and comfy seats, while other more rough-and-ready types work best at the bottom of the garden and function much like a compost heap, in that you mix the dry matter in with an active heap. Quick-growing plants, such as willow, are also planted around the site to soak up excess moisture. For this reason, these systems were known as 'tree bogs' in their early years.

Like any system of composting, the heat generated by the bacteria present in a compost loo is enough to kill off pathogens and make the resulting compost safe to use. The key thing to understand is that our own waste matter is some 90 per cent water, with the rest organic matter. In a dry system, most of this water will evaporate quickly. (In a conventional water-flushing system, you are mixing human effluent with water and creating much more volume to deal with.) Also the new waste is added to compost that is already active and has a good air flow to ensure that the process works quickly. At the end of the cycle, there is a safe, odourless product.

COMPOSTING

'Night soil', or at least the composted version, is at the top of the pyramid for recycling, but the composting of other household and garden wastes also presents a wonderful opportunity to return to nature some of her generously given bounty. I have talked about composting green waste from the kitchen and garden in *Chapter 2: Growing*, but many waste products from the house need to be collected and taken off-site or reused. Newspapers and cardboard and paper packaging are an enormous concern, as they are bulky. You can take them to the municipal recycling centre but, if shredded, they will compost well at home or they will make good materials for animal bedding. Cardboard also works as a good mulch in the garden for retaining moisture and suppressing weeds. Also, the wood ash from a house fire or outdoor bonfire offers a good source of potash that can be used on fruit and vegetable growing beds. Potash is a vital element for plant growth, but is highly soluble in water and thus tends to wash out of the soil, so regular application is necessary.

THE HARD STUFF

Beyond biodegradable matter, what do we do with the glass, the metal, the plastic, the tyres and all the other material that accumulates and has no obvious home? The answer is recycling. Glass goes to make more glass, the tin cans get turned into wire, plastic becomes yet more packaging and so on, all using far less energy to recycle than it costs to make them in the first place. So it is worth it. Also, before you send these materials off site, make sure that you don't have a use for them around the home or the holding.

CONCLUSION

CONCLUSION

My own practice of the concept of 'sufficient' has served me well, but the important thing is that everyone must find their own way of working with nature; there are no hard-and-fast rules. This intuitive approach of working reminds me of a Cornish dairy farmer I once visited, who sold unpasteurised milk from his herd of beautiful toffee-coloured Guernseys that grazed the herb-rich pastures of his west Cornwall farm. The farmer had no sets of figures or statistics about nutrition to show me, for either his cows or the people who drank the milk. He was so in tune with the natural cycles of life that he knew precisely what was good for his cows and therefore that the product that they brought forth was intrinsically good. His fundamental understanding of nature was what held him in good stead, and for those of us who desire to be more 'sufficient' in our lives, this is the first step.

Even for those without a garden, the aim should be to try and connect with nature on any level we can. We must start by trying to understand how our food is produced, where and by whom and under what conditions. Food is the most important thing for our health and well-being, so this book is primarily about the home production of it. Having been an organic grower, I champion this method of cultivation because I feel that the intention behind it is one of respect for the soil and nature. But this is not to say that conventional farmers do not have these ideals too. Yes, there are dust bowls in the US where once there were farms because of over-applications of artificial fertilisers but, on the whole, farmers have a clearer appreciation of how their soils and animals work than you or I. That the agrochemicals lobby, where no more than a handful of companies control almost three quarters of the world's pesticide sales, have a big hand in farming is no surprise. Thankfully, this is beginning to change as farmers begin to diversify and the demand for organic produce continues to grow.

We are by no means caught in a destructive spiral of chemical use. It has only been since the Second World War that wholesale chemical use came into play, and there will be plenty of incentive to reverse this use as the energy costs required to produce and apply these chemicals soar and pests and disease adapt so that the chemicals become less effective. The worry is that the chemicals will be replaced by genetically modified organisms, and although these are so far advanced that there may be no way back from their wholesale use, I am

Right: Our daughter, Rose Petherick, in the orchard at Cholwell Farm in South Devon, UK, where we are now practising 'Sufficient'.

certain that nature will find her way around those too. It was only 80 years ago that the world's farms were managed organically without chemicals. I believe that farming will be chemical-free again because the experiment with agrochemicals has been nothing short of a catastrophic failure.

So the message is to think carefully when choosing your food. Think seasonal, think local and, if you are not buying organic, try and work out which produce are likely to have been subjected to less chemical treatment than others. Investigate the possibility of buying garden produce and, please, buy from local growers if at all possible. These are people you will be able to connect with, and, if you buy from a supermarket – despite labelling – you will still be in the dark. Those of us in towns and cities may have to rely on vegetable box schemes to deliver or small growers on the outskirts of towns, but they are out there; your local food directory is a good place to start.

A change of attitude towards food, which recognises the importance of its provenance and an understanding of how it is produced, is the beginning of a journey of realisation. It will lead you to such causes as the Slow Food Movement and ultimately to a reversal in the way you think food should be grown and how we farm. In his excellent book *The Farm*, Richard Benson describes how, in Britain, the family farm was a mixture of arable and livestock farming created by fencing and hedging small parcels of land as part of the Enclosure Acts of the eighteenth and nineteenth centuries. Its purpose was to combine livestock and arable so that if one crop failed, there was another as back-up. You could feed crops to the animals and then put this back into the land by fertilising it with animal manure. The loss of such mixed farms in favour of monocultures, are a case of economies of scale and money defying common sense and 'that jarring, perplexing fact of the modern world: that the logic and geography of business is not syncopated with the logic of human feelings'. Harrowing but true. Thankfully, the reversal is under way, and while largescale conventional farming may be with us for a while yet, there is huge clamour for change back to the old system.

DO-IT-YOURSELF

A good way to start this process of change is to begin growing your own food. The metamorphosis that happens to people (who don't see themselves as gardeners and then suddenly give it a go) is astonishing. I would urge that you have a crack at it because it is good fun. It will also connect you with nature, in ways that you never imagined, on whatever scale you practice it.

When you do get started, it is important to have a plan for the layout of the productive garden and understand how the weather affects it – wind, rain and sun. You must also sort out basic services to the site, such as water and drainage, because it is a terrible sweat to have to worry about these after the event. The last thing you want is to have to dig up the garden to lay a pipe when it is planted, looking beautiful and starting to become productive. Intricate design is up to the individual. My own feeling is that what suits you is what is best as long as it is handy and you are comfortable with it. There is no one set of rules. There are energy-saving methods and good advice everywhere. I feel, though, that some exponents of various theories have become too rigid in their desire to follow their adopted principles, to the point where they become obsessive and immovable. There is no one way but yours; you know your house, garden and soil. You also know what you like and what is going to make you happy, so stick with that.

And why not dig up the front lawn and give it over to vegetables? You'd be the first. I don't get that many looks at back gardens, but I make a point of peering nosily at front gardens when I'm passing by – I am a gardener, that's what we do – but I don't see many front gardens given over to fruit cages or rows of lettuces. Summer bedding yes, but not edibles. It's time we got over this absurd stigma that edible plants are workmanlike and should be out of view. Show them off and be proud. But what about exhaust fumes? Front gardens are often closer to the road, but these fumes are no worse than insecticide. Anyway, you don't have to grow lettuces by the front door. Why not giant globe artichokes or even Jerusalem artichokes? They will look stupendous.

CHOICES

This is the world we live in, the one of multiple choices. It serves us well in some spheres, but it can be the devil in others. Growing is no different from anything else in this respect. When it comes to choosing what to grow, the choices are many so you must narrow them down. The more you grow, the less time you will have to look after it and, believe me, this is a time-consuming business. So keep it simple to begin with and don't try and grow too much. Deal with as many of the 'problems' that you discover in your garden as early as possible. It may even be worth sacrificing a season's crop growing to deal

with a specific weed problem. You can't possibly hope to achieve all that much in a single year, so easy does it. Set aside areas for certain crops, get the perennials in early and remember to work to the rotation because by this method you will properly build your soil fertility.

The compost side of things is absolutely critical and is a good place for all the household biodegradable foodstuffs rather than the dustbin. As I have explained earlier (see pages 80–83), there is a lot of nonsense talked about compost. It is not rocket science. If you just chuck the garden and kitchen waste in a heap, it will rot down, quicker if the matter is small, but much, much slower if it is in big chunks. It may not seem worth the effort to make a trip to the compost heap with only small quantities of fruit or vegetable peelings, but believe me it is, and you won't regret it. Like growing plants successfully, you will find that once you have made a batch of good compost you will always want to make it. This compost makes the perfect base for any plant grown in a container, and it will save trips to the garden centre to buy compost in bags.

I have tried to dissuade the reader from looking at the garden as a commercial venture but I feel, in retrospect, that perhaps I have been a little hasty on this subject. In Rodale's *Growing Fruit and Vegetables by the Organic Method*– the American must-have book on organics and the first of its kind – there is a section devoted to growing fruit and vegetables to sell. It is adorned by pictures of immaculate roadside stalls dripping with gigantic watermelons, juicy peaches and bunches of sweet grapes, just as the Californian growers shown are nice, young, clean-cut boys with tidy hair and flashing white teeth. Odd, considering it was the 1960s! What I am driving at is that serious commercial growing is a full-time job but, if you have excess produce, there is absolutely no harm in trying to sell it or barter. That roadside stall will soon make you a lot of friends, especially if you have bunches of sweet peas and fresh fruit at competitive prices. It's a good thing to do if you have the time and feel like interacting with buyers. There are no legal restrictions and a table with some freshly cut flowers, plants, or fruit or vegetable produce on it will not get you into any trouble with the law, although their are DEFRA regulations when it comes to selling eggs, dairy produce and meat.

As soon as you have crops growing in your garden, you are well under way. What happens next is up to you, and whether or not you are a gardener who is happy to let plants grow together with weeds or you prefer neat rows, you

Left: Whatever you can produce will be worth it and any effort in this direction will help towards the cause of 'sufficient'.

will have to garden even if it is simply to sow and harvest. This is where attention to detail comes in. Observation is everything. You must look and learn and try to figure out what is happening and why. I have skirted around the issue of pest and disease control because very often there are no clear-cut answers. Some problems, usually caused by a third party such as rabbits, pigeons, or slugs, will make themselves known quite quickly. Prevention is always better than cure because it means you can catch the problem before it gets out of hand. Cure means that you will have to reverse a trend, and that is almost always more difficult. The true 'sufficient' outlook is not to expect or demand too much from your garden. Again, easy does it; there is always a successional crop to plant a few weeks later that might do better, or even next year to find the balance that the garden.

Although I set great store by attention to detail, there is no greater teacher than nature. Plants want to grow and will find their own way given the resources that nature provides freely – light, air, water and food. Our job as growers is to help this process along the way, whether by the use of mulches, liquid feeds, correct spacing and layout, or structures and nets to help plants climb and protect them from pest attack.

When it comes to rearing animals, things are slightly different because most animals do not look after themselves. For this, more time and devotion to duty is required. Plants and animals work well together and provide for each other and us. But again, start small and don't overburden yourself. I hope, though, that this book will encourage you to take up the challenge, for there is much to be gained.

Perhaps the most important aspect of 'sufficient' and all that it brings is the one that is the least tangible. The slightest effort in this direction can change consciousness on a very deep level. If you put out an intention to change, what follows is that the ripples will flood out through whatever you want them to. Nature on the surface is simple; if left alone, she will do the right thing, just as the forest finds the right balance. This is something that we discover ourselves as we tread the path of 'sufficient'. We also need to find this out for ourselves in all other areas of life.

FURTHER READING

An Agricultural Testament, Sir Albert Howard, Oxford University Press, 1940

Designing and Maintaining Your Edible Landscape Naturally, Michael Kourik, Metamorphic Press, 1986

Ecovillages: New Frontiers for Sustainability, Jonathan Dawson, Green Books, 2006

Farms of Tomorrow Revisited: Communities Supporting Farms, Farms Supporting Communities, Steve Mc Fadden and Trauger Groh, Chelsea Green, 1997

Farms of Tomorrow, Trauger Groh, Biodynamic Farming and Gardening Association, 1990

Forest Farming, Robert Hart and Sholto Douglas, ITDG Publishing, 1984

Forest Gardening, Robert Hart, Green Earth Books, 1996

Grow Your Own Fruit and Vegetables, Lawrence D. Hills, Faber and Faber, 1973

How to Grow Vegetables and Fruits by the Organic Method, J.I. Rodale, Rodale Press, 1961

Nourishing Traditions, Sally Fallon, New Century Press, 1998

Permaculture, Bill Mollison, Eco-Logic Books/Worldly Goods, 1990

Plants for a Future, Ken Fern, Permanent Publications, 1997

Resurgence magazine, see www.resurgence.org

Silent Spring, Rachel Carson, Penguin, 1962

The Economics of Climate Change – The Stern Review, Nicholas Stern, Cabinet Office HM Treasury, Cambridge University Press, 2007

The Farm, Richard Benson, Penguin, 2005

The Four Season Harvest, Elliot Coleman, Chelsea Green, 1999

The Generous Earth, Philip Oyler, Hodder & Stoughton, 1950

The Green Fool, Patrick Kavanagh, Penguin, 1938

The Natural Way of Farming, Masanobu Fukuoka, Other India Press, 1985

The New Complete Book of Self-Sufficiency, John Seymour, Dorling Kindersley, 2003

The One-Straw Revolution: Introduction to Natural Farming (Originally published in Japan as Shizen Noho Wara Ippon No Kakumei), Masanobu Fukuoka, Other India Press, new edition 1992

Utopian Dreams: In Search of a Better Life, Tobias Jones, Faber and Faber, 2007

We Want Real Food, Graham Harvey, Constable, 2006

INDEX

PICTURE CREDITS

Melanie Eclare: pages 11, 53, 115, 119, 125, 128, 137, 140, 145, 148-149, 166, 200, 209, 216, 223, 265, 275

Francesca Yorke: pages 14, 58, 61, 63, 83, 84, 99, 111, 121, 132, 134, 139, 143, 146, 153, 159, 165, 170, 175, 181, 205, 224, 227, 232, 249, 278

Alamy Images: pages 22L Mark Bolton, 22R AGStock USA, Inc., 27 Eye Ubiquitous, 29 Steven May, 32 Andrew Duke, 33 Andrew Duke, 46 Skyscan Photolibrary, 47 Elizabeth Whiting & Assiciates, 69 Paul Bradforth, 87 Travel Ink, 173 The Garden Picture Library, 191 Andrew Paterson, 197 Kevin Howchin, 221 Kim Karpeles, 241 BL Images Ltd, 243 Photofusion Picture Library, 244 Jeff Morgan Alternative Technology, 247 Edward Parker, 253 Keith M Law, 257 Paul Glendell, 262 Andrew Butterton, 266 Angus Mackie

Corbis: page 42 Free Agents Limited

ACKNOWLEDGEMENTS

The concept of 'Sufficient' is very real for me but the actual word came in a blinding flash of inspiration from Polly Powell, who commissioned this book. So my first acknowledgment is to Polly whose coining of the word 'sufficient' is right in line with the thinking of the day – thanks to you for allowing me free rein to write about the things that matter to me. This is not something that everyone gets to do in their lifetime, so I am extremely grateful – it has been great fun.

Everyone at Pavilion Books has been very relaxed and easy-going so the process could not have been easier. Thank you so much Emily Preece-Morrison, my editor-in-chief at Pavilion, for making every step enjoyable. Wise guidance has come at all times, allied with freedom to expand, so thank you Emily – writing is a joy with such friends to hand.

Also a big thank you to Bella Pringle, who edited the book and who set me right on many occasions, trawling through text which at times must have presented a proper challenge.

Lotte Oldfield's illustrations are wonderful and really make the book – the whole look is spot on. Thanks are also due to Lotte's sister, Kate Oldfield, who first took up the possibility of a book from me on growing and organics in general.

Although *Sufficient* is a text-led book there are some beautiful pictures, mostly taken by my wife Melanie who is always there to pick up the pieces and come up with a jaw-dropping image – so thank you Darling. And thanks also to Francesca Yorke who supplied many of the other images.

Lots more thanks, including to Angie Dodd from Seriously Good in Cornwall whose Slow Food Database was a big help. To the staff and children of the South Devon Steiner School and Joe Clarke for letting us take pictures of the incredible garden at Hood Manor. And to Derek Lapworth of Velwell Orchard, who let us photograph at his bio-dynamic orchard which adjoins our own farm at Cholwell in Devon and whose vision is at all times inspirational.

Thanks as ever to our agent Jane Turnbull for believing in the book and working so hard to get it away with Pavilion.

And finally, thanks to all those small gardeners and growers who keep the dream alive and who will be the future of food security.

This book was printed on 130gsm Munken Pure paper, produced by Arctic Paper Munkedals AB, which has a long tradition of environmental responsibilty. Papers are created to minimize the impact on the environment. For example, they have not been subject to chlorine bleaching, pulp is taken from sustainable forests FSC, and production processes have been developed to reduce harmful effects and effluents.

First published in the United Kingdom in 2007 by
Pavilion Books
Old West London Magistrates Court
10 Southcombe Street
London
W14 0RA

An imprint of Anova Books Company Ltd

Design and layout © Pavilion, 2007
Text © Tom Petherick, 2007
Photography © see picture credits

Senior Editor: Emily Preece-Morrison
Designer: Louise Leffler @ Sticks Design
Photographers: Melanie Eclare and Francesca Yorke, except where otherwise credited
Illustrator: Lotte Oldfield
Copy Editor: Bella Pringle
Proofreader: Caroline Curtis
Indexer: Helen Snaith
Production: Alexia Turner

ISBN 9781862057739

A CIP catalogue record for this book is available from the British Library.

10 9 8 7 6 5 4 3 2 1

Reproduction by Mission Productions, Hong Kong
Printed and bound by SNP Leefung, China

www.anovabooks.com